Tanya Mincheva

Kinematics and Dynamics of Generalized-Symetric Sets

Tanya Mincheva

Kinematics and Dynamics of Generalized-Symetric Sets

Applications in Number Theory: Theorem of Goldbach and Riemann's Hypothesis

LAP LAMBERT Academic Publishing

Impressum / Imprint
Bibliografische Information der Deutschen Nationalbibliothek: Die Deutsche Nationalbibliothek verzeichnet diese Publikation in der Deutschen Nationalbibliografie; detaillierte bibliografische Daten sind im Internet über http://dnb.d-nb.de abrufbar.
Alle in diesem Buch genannten Marken und Produktnamen unterliegen warenzeichen-, marken- oder patentrechtlichem Schutz bzw. sind Warenzeichen oder eingetragene Warenzeichen der jeweiligen Inhaber. Die Wiedergabe von Marken, Produktnamen, Gebrauchsnamen, Handelsnamen, Warenbezeichnungen u.s.w. in diesem Werk berechtigt auch ohne besondere Kennzeichnung nicht zu der Annahme, dass solche Namen im Sinne der Warenzeichen- und Markenschutzgesetzgebung als frei zu betrachten wären und daher von jedermann benutzt werden dürften.

Bibliographic information published by the Deutsche Nationalbibliothek: The Deutsche Nationalbibliothek lists this publication in the Deutsche Nationalbibliografie; detailed bibliographic data are available in the Internet at http://dnb.d-nb.de.
Any brand names and product names mentioned in this book are subject to trademark, brand or patent protection and are trademarks or registered trademarks of their respective holders. The use of brand names, product names, common names, trade names, product descriptions etc. even without a particular marking in this works is in no way to be construed to mean that such names may be regarded as unrestricted in respect of trademark and brand protection legislation and could thus be used by anyone.

Coverbild / Cover image: www.ingimage.com

Verlag / Publisher:
LAP LAMBERT Academic Publishing
ist ein Imprint der / is a trademark of
OmniScriptum GmbH & Co. KG
Heinrich-Böcking-Str. 6-8, 66121 Saarbrücken, Deutschland / Germany
Email: info@lap-publishing.com

Herstellung: siehe letzte Seite /
Printed at: see last page
ISBN: 978-3-659-21882-8

Tanya Kirilova Mincheva

KINEMATICS AND DYNAMICS OF
GENERALIZET-SYMETRIC SETS

APPLICATIONS IN NUMBER THEORY:
THEREM OF GOLDBACH AND RIEMANN'S HYPOTESIS

March 2014, Sofia-Bulgaria

CONTENT

Tanya Mincheva

KINEMATICS AND DYNAMICS OF GENERALIZED-SYMETRIC SETS AND

THEIR APPLICATIONS IN NUMBER THEORY

Tanya Kirilova Mincheva
Sofia-Bulgaria
E-mail: tmintcheva@yahoo.com

Abstract. The definition of "**arithmetic progression**" is viewed as a generalization of the concept of symmetry sets on the real axis. We use the positive whole numbers. That interpretation is used to prove the theorem of Goldbach.
Due to the interest in the topic by people who are not mathematicians we will try to be more detailed in some places.
Let us remember the Ox-axis.
The point O is the center of symmetry. The sum of any two opposite numbers is zero. If we consider a finite interval $[-n \ n]$, then opposite numbers are numbers that are equally spaced by the finite elements.

Definition: In mathematics an arithmetic progression (AP) or an arithmetic sequence is a sequence of numbers such that the difference between the consecutive terms is constant If the initial term of an arithmetic progression is a_1 and the common difference of successive members is d, then the nth term of the sequence $\{a_n\}_{n=1}^{\infty}$ is given by:

$$a_n = a_1 + (n-1)d$$

For our study is most important the following property of any finite arithmetic progression. **Basic property: The sums of all numbers, equidistant from the endmost elements, are the same.**
This is our fundamental and leading thought in that research.

Theorem. Each finite arithmetic progression can be viewed as a generalization of the definition for symmetry.
Proof.
The proof is elementary.
Let

$-n, -(n-1), -(n-2),\ldots, -3, -2, -1, 0, 1, 2, 3, \ldots n-2, n-1, n; n \in N$

be one example of finite symmetrical multitude on Ox. Obviously this is one arithmetic progression, where
$a_1 = -n$, $d = 1$ and from our fundamental thought

$$-n + n = (-(n-1)) + n-1 = (-(n-2)) + n-2 = \ldots = -2+2 = -1+1 = 2.0 = p = 0$$

For any $\div\ a_1, a_2, a_3, \ldots a_{n-2}, a_{n-1}, a_n$ too. Namely:

3

$a_1 + a_n = a_2 + a_{n-1} = a_3 + a_{n-2} = \dots = p$,

Where $p = 2a_k$, when $n = 2k+1$ and $p = a_k + a_{k+1}$, when $n = 2k$

The only difference is in the values of p. That finishes the proof.

For that reason each finite arithmetic progression we call **generalized symmetrical multitude**

Examples:

I. $-5, -4, -3, -2, -1, 0, 1, 2, 3, 4, 5.$ $\Rightarrow$

$\Rightarrow -5+5 = -4+4 = -3+3 = -2+2 = -1+1 = 2.0 \, ; p = 0.$

II. $1, 2, 3.$ $\Rightarrow$

$\Rightarrow 1+3 = 2.2. \quad p = 4.$

III. $1, 2, 3, 4, 5.$ $\Rightarrow$

$\Rightarrow 1+5 = 2+4 = 2.3 \Rightarrow p = 6.$

IV. $1, 2, 3, 4, 5, 6, 7.$ $\Rightarrow$

$\Rightarrow 1+7 = 2+6 = 3+5 = 2.4 \Rightarrow p = 8.$

Although these examples are the most elementary, our ideas for their consideration should be noted. All progressions are finite; contain only odd number of members and defining parameters for all are (except the first example) $a_1 = 1, d = 1$.

These progressions and all others we will write in the following way, using

The basic property

$2.2 = 4$ for $= P_3 = \div 1, 2, 3.$, $n = 3.$

$1+3$

$2+2$

$2.3 = 6$ for $P_5 = = \div 1, 2, 3, 4, 5. \, a_1 = 1, d = 1, n = 5.$

$1+5$

$2+4$

$3+3$

$2.4 = 8$ for $P_7 = \div 1, 2, 3, 4, 5, 6, 7. \, a_1 = 1, d = 1, n = 7.$

$1+7$

$2+6$

$3+5$

$4+4$

2.5=10   for $P_9 = \div 1, 2, 3, 4, 5, 6, 7, 8, 9.$ $a_1 = 1, d = 1$, $n = 9$
1+9
2+8
3+7
4+6
5+5

2.6=12 for $P_{11} = \div 1, 2, 3, 4, 5, 6, 7, 8, 9, 10, 11.$ $a_1 = 1, d = 1$, $n=11.$
1+11
2+10
3+9
4+8
5+7
6+6

2.7=14   for $P_{13} = \div 1,2,3,4,5,6,7,8,9,10,11,12,13.$ $a_1 = 1, d = 1$, $n=13.$
1+13
2+12
3+11
4+10
5+9
6+8
7+7
....

Therefore, we can write a sequence, the elements of which are multitudes, namely arithmetic progressions where $a_1 = 1, d = 1$

$$\{P\}_{n=3}^{\infty} = \div P_3, \div P_5, \div P_7, \ldots, \div P_{2n+1}, \ldots$$

For our research are important also the following progressions:

$A_1 = \div 1$ $a_1 = 1, d = 0$, n=1.

$A_2 = \div 1, 3.$ $a_1 = 1, d = 2$; n=2.
        2.2=4
        1+3

$A_3 = \div 1, 3, 5.$ $a_1 = 1, d = 2$; n=3.
        2.3=6
        1+5
        3+3

$A_4 = \div$ 1, 3, 5, 7. $a_1 = 1, d = 2$; n=4.
    2.4=8
    1+7
    3+5

$A_5 = \div$ 1, 3, 5, 7, 9. $a_1 = 1, d = 2$; n=5.
    2.5=10
    1+9
    3+7
    5+5

$A_6 = \div$ 1, 3, 5, 7, 9, 11. $a_1 = 1, d = 2$; n=6.
    2.6=12
    1+11
    3+9
    5+7

$A_7 = \div$ 1, 3, 5, 7, 9, 11, 13. $a_1 = 1, d = 2$; n=7
    2.7=14
    1+13
    3+11
    5+9
    7+7

    ...   ...
    ...   ...

$A_{n-1} = \div 1, 3, 5... (2(n-1)-1)$ $a_1 = 1, d = 2$; $\rightarrow$ (n-1).

$A_n = \div 1, 3, 5... (2n-1)$ $a_1 = 1, d = 2$; $\rightarrow$ n

$A_{n+1} = \div 1, 3, 5... (2(n+1)-1)$ $a_1 = 1, d = 2$; $\rightarrow$ (n+1)
...
...

By analogy, we can write another sequence, the elements of which are multitudes too. Namely, this is the next row

$$\{A_n\}_{n=2}^{\infty} = \div A_2, \div A_3, \; ... \div A_n, ...$$

In which the elements are sets of arithmetic progressions with $a_1 = 1, d = 2$. The subscript represents the number of elements in the marked progressions.
Arithmetic progressions are very well studied.

Fundamental difference between progressions $\{A_n\}_{n=2}^{\infty}$ and $\{P\}_{n=3}^{\infty}$ does not exist. In both cases after application of the basic property are constructed the sets of all positive even integers.

In the second group progressions of even quantity of numbers are represented by sums of odd numbers. All prime numbers are odd. For this reason, we will work with progressions $\{\div A_n\}$ $n = 2,3,...$ where

$$a_1 = 1, d = 2$$

We will introduce each of the elements of the last progression in a suitable matrix form. We think this is a fair way.

The plus sign is not written. It is understood. That implied:

$A_2 = \div 1, 3. \Rightarrow$

$$4$$
$$\begin{pmatrix} 1 & 3* \\ 2*2* \end{pmatrix}$$

$A_3 = \div 1, 3, 5.$

$$6$$
$$\begin{pmatrix} 1 & 5* \\ 3* & 3* \end{pmatrix}$$

$A_4 = \div 1, 3, 5, 7.$

$$8$$
$$\begin{pmatrix} 1 & 7* \\ 3* & 5* \end{pmatrix}$$

$A_5 = \div 1, 3, 5, 7, 9.$

$$10$$
$$\begin{pmatrix} 1 & 9 \\ 3* & 7* \\ 5* & 5* \end{pmatrix}$$

$A_6 = \div 1, 3, 5, 7, 9, 11.$

$$12$$
$$\begin{pmatrix} 1 & 11* \\ 3* & 9 \\ 5* & 7* \end{pmatrix}$$

$A_7 = \div$ 1, 3, 5, 7, 9, 11, 13.

$$14$$

$$\begin{pmatrix} 1 & 13* \\ 3* & 11* \\ 5* & 9 \\ 7* & 7* \end{pmatrix}$$

$A_8 = \div$ 1, 3, 5, 7, 9, 11, 13, 15.

$$16$$

$$\begin{pmatrix} 1 & 15 \\ 3* & 13* \\ 5* & 11* \\ 7* & 9 \end{pmatrix}$$

$A_9 = \div$ 1, 3, 5, 7, 9, 11, 13, 15, 17.

$$18$$

$$\begin{pmatrix} 1 & 17 \\ 3* & 15 \\ 5* & 13* \\ 7* & 11* \\ 9 & 9 \end{pmatrix}$$

$A_{10} = \div$ 1, 3, 5, 7, 9, 11, 13, 15, 17, 19

$$20$$

$$\begin{pmatrix} 1 & 19* \\ 3* & 17* \\ 5* & 15 \\ 7* & 13* \\ 9 & 11* \end{pmatrix}$$

$A_{11} = \div$ 1, 3, 5, 7, 9, 11, 13, 15, 17, 19, 21.

$$22$$

$$\begin{pmatrix} 1 & 21 \\ 3* & 19* \\ 5* & 17* \\ 7* & 15 \\ 9 & 13* \\ 11* & 11* \end{pmatrix}$$

$A_{12} = \div 1, 3, 5, 7, 9, 11, 13, 15, 17, 19, 21, 23$ 24

$$\begin{pmatrix} 1 & 23* \\ 3* & 21 \\ 5* & 19* \\ 7* & 17* \\ 9 & 15 \\ 11* & 13* \end{pmatrix}$$

$A_{13} = \div 1, 3, 5, 7, 9, 11, 13, 15, 17, 19, 21, 23, 25.$ 26

$$\begin{pmatrix} 1 & 25 \\ 3* & 23* \\ 5* & 21 \\ 7* & 19* \\ 9 & 17* \\ 11* & 15 \\ 13* & 13* \end{pmatrix}$$

$A_{14} = \div 1, 3, 5, 7, 9, 11, 13, 15, 17, 19, 21, 23, 25, 27$ 28

$$\begin{pmatrix} 1 & 27 \\ 3* & 25 \\ 5* & 23* \\ 7* & 21 \\ 9 & 19* \\ 11* & 17* \\ 13* & 15 \end{pmatrix}$$

$A_{15} = \div 1, 3, 5, 7, 9, 11, 13, 15, 17, 19, 21, 23, 25, 27, 29$ 30

$$\begin{pmatrix} 1 & 29* \\ 3* & 27 \\ 5* & 25 \\ 7* & 23* \\ 9 & 21 \\ 11* & 19* \\ 13* & 17* \\ 15 & 15 \end{pmatrix}$$

The stars written as superscripts denote primes. The case of A_2 is an exception, but it is elementary.

Lemma.

The row $1,3,...,1+2n$ gives all the possibilities in which every positive even numbers can be expressed as sums of odd numbers.

Proof.

The proof is elementary - with mathematical induction.

For A_2: 1+3=2.2=4, for A_3: 1+5=2.3=6 this is true. Let's assume that for the progression

$$A_n = \div\, 1, 3, 5,..., 2n\text{-}5, 2n\text{-}3, 2n\text{-}1,$$

whit n members the lemma is true too

I.e. 1+2n-1=3+2n-3=…=2n,

Then for the next progression   1, 3, 5… 2(n+1)-5, 2(n+1)-3, 2(n+1)-1, we have

1+2(n+1)-1=3+2(n+1)-3=…=2(n+1).
This finishes the proof.

Remark: for $n=2k$ we have 1+2.2 k-1=…=2.2k-1 +2.2k+1 =.4k.

For $n=2k+1$ analogy 1+2(2k+1)-1=…=2(2k+1)
In both cases we get even numbers.

$$\div\ 1, 3, 5, ..., 2n\text{-}1. \qquad \text{Or}$$

$$\div\ 1, 3, 5, ..., 2n\text{-}1, 2n\text{+}1.$$

Definition. Let us have three arithmetic progressions.(see fig.1)

$$2n=2(2k+1) \qquad 2(n+1)=2(2k+2) \qquad 2(n+2)=2(2k+3)$$

$$A_{2k+1}=\begin{pmatrix} 1 & 4k+1 \\ 3 & 4k-1 \\ 5 & 4k-3 \\ ... \\ ... \\ 2k-1 & 2k+3 \\ 2k+1 & 2k+1 \end{pmatrix},\ A_{2k+2}=\begin{pmatrix} 1 & 4k+3 \\ 3 & 4k+1 \\ 5 & 4k-1 \\ ... \\ ... \\ 2k-1 & 2k+5 \\ 2k+1 & 2k+3 \end{pmatrix},\ A_{2k+3}=\begin{pmatrix} 1 & 4k+5 \\ 3 & 4k+3 \\ 5 & 4k+1 \\ ... \\ ... \\ 2k-1 & 2k+7 \\ 2k+1 & 2k+5 \\ 2k+3 & 2k+3 \end{pmatrix}$$

Fig.1

For A_{2k+1} the conformity: $3 \to 4k+1$; $5 \to 4k-1$; ..., $2k+1 \to 2k+3$ and

for A_{2k+3} the conformity: $1 \to 4k+3$; $3 \to 4k+1$; ..., $2k+1 \to 2k+3$.

we call **internal dependence or internal dynamics**. Arrows indicate how the previous or following matrix for every $n \in N$ must be formed.
The arrows indicate internal INTERACTION or internal DYNAMICS through which it is possible a TRANSITION -KINEMATICS toward the next matrix.
This definition is an analogue of the formula for the relationship between two consecutive terms in an arithmetic progression.
I think this idea needs to be clarified with many examples, because it is very important for our proof. For this reason, we already wrote finite numerical progressions. Now we'll use them for our purpose.
Let us consider the matrices for the numbers 26, 28 and 30.

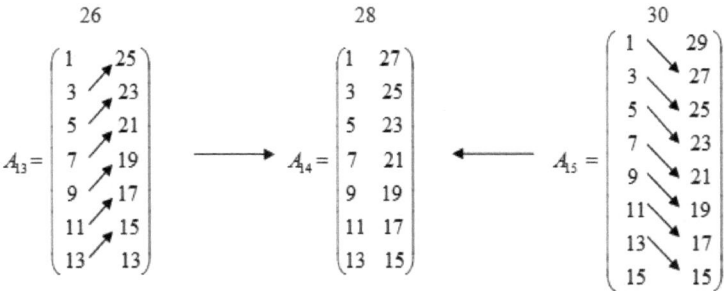

Fig.2

In using these matrices, we can write the members of the progressions in a new way. In each row of these matrices are members which are equidistant from the end elements, i.e. with equal sums. If the progression contains an odd number of members (for example 13 or 15 or 17), in this case, in the last line of the matrix the middle element is written twice. But

(1) $(3; 25), (5; 23), (7; 21), (9; 19), (11; 17), (13; 15) \subset A_{14}$

It is understood that the arrows in the matrix for A_{13} represent the elements in A_{14}.
Analogically, the inverse dynamic from A_{15} to A_{14}. I.e.

(2) $(1: 27), (3; 25), (5, 23), (7; 21), (9; 19), (11; 17), (13; 15)$. Or $A_{15} \supset A_{14}$

Before we begin the proof we will make the following remarks:
I. Formula (1) we can make the transition from A_{13} to progression A_{14}.
II. Similarly, formula (2) in transition to the previous progression.

11

Theorem of Goldbach (1690-1764). Every even number greater than two can by written as the sum of two primes.

Generally, for three consecutive arithmetic progressions, we see immediately:

I. All members in the left column in the transition to the next matrix are stored.

II. Similarly, for the right columns.

Remark. We have already said that people who are not mathematicians are also interested by this theorem. In the beginning, we will indicate the idea of the proof of the Theorem of Goldbach with numerical examples. We consider

$$
\div A_{13} = \begin{pmatrix} 1 & 25 \\ 3* & 23* \\ 5* & 21 \\ 7* & 19* \\ 9 & 17* \\ 11* & 15 \\ 13* & 13* \end{pmatrix}, \quad
\div A_{14} = \begin{pmatrix} 1 & 27 \\ 3* & 25 \\ 5* & 23* \\ 7* & 21 \\ 9 & 19* \\ 11* & 17* \\ 13* & 15 \end{pmatrix}, \quad
\div A_{15} = \begin{pmatrix} 1 & 29* \\ 3* & 27 \\ 5* & 25 \\ 7* & 23* \\ 9 & 21 \\ 11* & 19* \\ 13* & 17* \\ 15 & 15 \end{pmatrix}, \quad
\div A_{16} = \begin{pmatrix} 1 & 31 \\ 3* & 29* \\ 5* & 27 \\ 7* & 25 \\ 9 & 23* \\ 11* & 21 \\ 13* & 19* \\ 15 & 17* \end{pmatrix}.
$$

(columns headed: 26, 28, 30, 32)

Immediately we assume that for all progressions to $\div A_i$, $i = 2, 3, 4 \ldots 14$ the theorem is true. We have to prove that and for the next A_{15} this is correct too. Let's assume that A_{15} is not correct. This means that the matrix A_{15} does not have a row with two primes. In the beginning we consider the elements of the left column of the A_{15}. According to our assumption the numbers 3*, 5*, 7*, 11*, 13* are prime, but 27, 25, 23*, 19*, 17* all are composite numbers. For two successive arithmetic progressions the difference in the number of rows is at most one.

For that reason, in the reverse transformation (internal dynamics in A_{15}), after we use as starting points the integers of the right column, we shall restore every row of A_{14} without a row from two primes. This is not possible. This is not possible because, according to our assumption in A_{14} there is a pair of prime numbers. I.e. in a A_{14} there is a number that is prime and composite at the same time. That is a contradiction. Namely 23 (to our assumption) as a member of A_{15} is composite number, but as member of A_{14} is prime.

By analogy, if first we start with the right column of A_{15}, in other words, when moving from to on one of the lines there are two primes. This denotes that 23, 19, 17 are prime, but 7, 11, 13 are not prime numbers. This is impossible, because in contrary motion to A_{14} there is a row of two primes. This means that there is a number which is prime number and at the same time is composite number.

We consider the left and right column separately. Obviously, they can be viewed simultaneously without restricting the strength of the evidence.

However, we think, these cases are sufficient to conclude a common method.

Proof of theorem of Goldbach.
Once we have introduced the notion of conditional symmetry and its matrix recording, we will prove the theorem by the method of complete mathematical induction.
Let

$$\div A_2, \div A_3, \div A_4, \ldots, \div A_n, \div A_{n+1}, \ldots$$

be a row in which all elements are multitudes - finite arithmetic progressions. In all progressions the defining parameters are:

$a_1 = 1$, $d = 2$, index n denotes the number of members and

$a_n = 1 + (n-1)2$, I.e. $a_n = 2n - 1$.

Namely, we already considered them. Or for a row of matrices we have

$$\begin{pmatrix} 1 & 3* \\ 2* & 2* \end{pmatrix}, \begin{pmatrix} 1 & 5* \\ 3* & 3* \end{pmatrix}, \begin{pmatrix} 1 & 7* \\ 3* & 5* \end{pmatrix}, \begin{pmatrix} 1 & 9 \\ 3* & 7* \\ 5* & 5* \end{pmatrix}, \ldots, \begin{pmatrix} 1 & 2n-1 \\ 3 & 2n-3 \\ \cdots \\ n-1 & n+1 \end{pmatrix}, \begin{pmatrix} 1 & 2(n+1)-1 \\ 3 & 2(n+1)-3 \\ \cdots \\ n & n+2 \end{pmatrix}, \ldots$$

Where

$$A_n = \begin{pmatrix} 1 & 2n-1 \\ 3 & 2n-3 \\ \cdots \\ n-1 & n+1 \end{pmatrix} \text{ and } A_{n+1} = \begin{pmatrix} 1 & 2(n+1)-1 \\ 3 & 2(n+1)-3 \\ \cdots \\ n & n+2 \end{pmatrix}$$

If we denote

A_n' is a matrix of numbers $\begin{pmatrix} 1 \\ 3 \\ \cdots \\ n-1 \end{pmatrix}$. Analogically for $A_n'' = \begin{pmatrix} 2n-1 \\ 2n-3 \\ \cdots \\ n+1 \end{pmatrix}$.

Then

$$A_{np}* = A_n' \cap^* A_n'' \, ; p < n.$$

The symbol $\cap^*$ refers to the formation of a matrix only by primes.

A_{np} is a matrix, in the rows of which there are prime numbers that belong both to A_n' and A_n''

Examples:

For $A_2 = \begin{pmatrix} 1 & 3* \\ 2* & 2* \end{pmatrix}$ $A_2' = \begin{pmatrix} 1 \\ 2* \end{pmatrix}$, $A_2'' = \begin{pmatrix} 3* \\ 2* \end{pmatrix} \rightarrow A_2' \cap^* A_2'' = (2* \quad 2*)$

For $A_3 = \begin{pmatrix} 1 & 5* \\ 3* & 3* \end{pmatrix}$ $A_3' = \begin{pmatrix} 1 \\ 3* \end{pmatrix}$, $A_3'' = \begin{pmatrix} 5* \\ 3* \end{pmatrix} \rightarrow A_3' \cap^* A_3'' = (3* \quad 3*)$

For $A_4 = \begin{pmatrix} 1 & 7* \\ 3* & 5* \end{pmatrix}$. $A_4' = \begin{pmatrix} 1 \\ 3* \end{pmatrix}$ $A_4'' = \begin{pmatrix} 7* \\ 5* \end{pmatrix} . \rightarrow A_4' \cap^* A_4'' = (3* \quad 5*)$

$A_5 = \begin{pmatrix} 1 & 9 \\ 3* & 7* \\ 5* & 5* \end{pmatrix}$ $A_5' = \begin{pmatrix} 1 \\ 3* \\ 5* \end{pmatrix}$, $A_5'' = \begin{pmatrix} 9 \\ 7* \\ 5* \end{pmatrix} \rightarrow A_5' \cap^* A_5'' = \begin{pmatrix} 3*7* \\ 5*5* \end{pmatrix}$

For all these initial cases we see that the theorem of Goldbach is true. Let for A_n

$$A_n = \begin{pmatrix} & 2n & \\ 1 & & 2n-1 \\ 3 & & 2n-3 \\ 5 & & 2n-5 \\ \cdots & & \\ k_1* & & 2n-k_1* \\ \cdots & & \\ k_2* & & 2n-k_2* \\ \cdots & & \\ \cdots & & \\ k_p* & & 2n-k_p* \\ \cdots & & \\ n-1 & & n+1 \end{pmatrix}$$

the theorem be true. We must prove that the theorem is true also for A_{n+1}.

If $n-1$ and $n+1$ are primes, as in the cases A_{12}, and A_{13} respectively, then the problem is solved, but this is misleading because this is not a characteristic of all elements from our row of finite arithmetic progressions. There also may be cases such as the A_9 and A_{10} etc. Let us compare the matrices A_n and A_{n+1}. The left column of A_{n+1} contains all the elements on the left column of A_n. Let us assume that after the transition in A_{n+1} there is not a single pair consisting of primes.

$$A_{n+1} = \begin{pmatrix} 1 & 2(n+1)-1 \\ 3 & 2(n+1)-3 \\ 5 & 2n-3 \\ \cdots\cdots \\ k_1 & 2(n+1)-k_1 \\ \cdots\cdots \\ k_2 & 2(n+1)-k_2 \\ \cdots\cdots \\ k_p & 2(n+1)-k_p \\ \cdots\cdots \\ n+1 & n+1 \end{pmatrix} = \begin{pmatrix} 1 & 2n+1 \\ 3 & 2n-1 \\ 5 & 2n-3 \\ \cdots\cdots \\ k_1 & 2n+2-k_1 \\ \cdots\cdots \\ k_2 & 2n+2-k_2 \\ \cdots\cdots \\ k_p & 2n+2-k_p \\ \cdots\cdots \\ n+1 & n+1 \end{pmatrix}, \text{ where } p<n.$$

Namely, we know that the left column is the same as in our assumption, but we suppose that in the entire matrix there aren't primes as elements of one row. Similarly, we use right column. But after reverse transformation we receive A_n. I.e. in A_n there is a number that is prime as element of A_n and composite in the same time as element of A_{n+1}. That is contradiction. This proves the theorem.

In conclusion, we present the other even numbers up to 120 as the sums of primes

$$n=32$$

$$A_{16} = \begin{pmatrix} 1 & 31 \\ 3* & 29* \\ 5* & 27 \\ 7* & 25 \\ 9 & 23* \\ 11* & 21 \\ 13* & 19* \\ 15 & 17* \end{pmatrix},$$

$$A_{16}{}' = \begin{pmatrix} 1 \\ 3* \\ 5* \\ 7* \\ 9 \\ 11* \\ 13* \\ 15 \end{pmatrix}, A_{16}{}'' = \begin{pmatrix} 31 \\ 29* \\ 27 \\ 25 \\ 23* \\ 21 \\ 19* \\ 17* \end{pmatrix}, A_{16}{}' \cap *A_{16}{}'' = \begin{pmatrix} 3*29* \\ 13*19* \end{pmatrix}$$

$2n=34$

$$A_{17} = \begin{pmatrix} 1 & 33 \\ 3* & 31* \\ 5* & 29* \\ 7* & 27 \\ 9 & 25 \\ 11* & 23* \\ 13* & 21 \\ 15 & 19* \\ 17* & 17* \end{pmatrix}, A_{17}{}' = \begin{pmatrix} 1 \\ 3* \\ 5* \\ 7* \\ 9 \\ 11* \\ 13* \\ 15 \\ 17* \end{pmatrix}, A_{17}{}'' = \begin{pmatrix} 33 \\ 31* \\ 29* \\ 27 \\ 25 \\ 23* \\ 21 \\ 19* \\ 17* \end{pmatrix} \rightarrow A_{17}* = A_{17}{}' \cap * A_{17}{}'' = \begin{pmatrix} 3* & 31* \\ 5* & 29* \\ 11* & 23* \\ 17* & 17* \end{pmatrix}$$

$2n=36$

$$A_{18} = \begin{pmatrix} 1 & 35 \\ 3* & 33 \\ 5* & 31* \\ 7* & 29* \\ 9 & 27 \\ 11* & 25 \\ 13* & 23* \\ 15 & 21 \\ 17* & 19* \end{pmatrix}, A_{18}{}' = \begin{pmatrix} 1 \\ 3* \\ 5* \\ 7* \\ 9 \\ 11* \\ 13* \\ 15 \\ 17* \end{pmatrix}, A_{18}{}'' = \begin{pmatrix} 35 \\ 33 \\ 31* \\ 29* \\ 27 \\ 25 \\ 23* \\ 21 \\ 19* \end{pmatrix} \rightarrow A_{18}* = A_{18}{}' \cap * A_{18}{}'' = \begin{pmatrix} 5* & 31* \\ 7* & 29* \\ 13* & 23* \\ 17* & 19* \end{pmatrix}$$

$2n=38$

$$A_{19} = \begin{pmatrix} 1 & 37* \\ 3* & 35 \\ 5* & 33 \\ 7* & 31* \\ 9 & 29* \\ 11* & 27 \\ 13* & 25 \\ 15 & 23* \\ 17* & 21 \\ 19* & 19* \end{pmatrix}, A_{19}{}' = \begin{pmatrix} 1 \\ 3* \\ 5* \\ 7* \\ 9 \\ 11* \\ 13* \\ 15 \\ 17* \\ 19* \end{pmatrix}, A_{19}{}'' = \begin{pmatrix} 37* \\ 35 \\ 33 \\ 31* \\ 29* \\ 27 \\ 25 \\ 23* \\ 21 \\ 19* \end{pmatrix} \rightarrow A_{19}* = A_{19}{}' \cap * A_{19}{}'' = \begin{pmatrix} 7* & 31* \\ 19* & 19* \end{pmatrix}$$

$2n=40$

$$A_{20} = \begin{pmatrix} 1 & 39 \\ 3* & 37* \\ 5* & 35 \\ 7* & 33 \\ 9 & 31* \\ 11* & 29* \\ 13* & 27 \\ 15 & 25 \\ 17* & 23* \\ 19* & 21 \end{pmatrix}, A_{20}{}' = \begin{pmatrix} 1 \\ 3* \\ 5* \\ 7* \\ 9 \\ 11* \\ 13* \\ 15 \\ 17* \\ 19* \end{pmatrix}, A_{20}{}'' = \begin{pmatrix} 39 \\ 37* \\ 35 \\ 33 \\ 31* \\ 29* \\ 27 \\ 25 \\ 23* \\ 21 \end{pmatrix}$$

$$\rightarrow A_{20}* = A_{20}{}' \cap * A_{20}{}'' = \begin{pmatrix} 3* & 37* \\ 11* & 29* \\ 17* & 23* \end{pmatrix}$$

$2n=42$

$$A_{21} = \begin{pmatrix} 1 & 41 \\ 3* & 39 \\ 5* & 37* \\ 7* & 35 \\ 9 & 33 \\ 11* & 31* \\ 13* & 29 \\ 15 & 27 \\ 17* & 25 \\ 19* & 23* \\ 21 & 21 \end{pmatrix}, A_{21}{}' = \begin{pmatrix} 1 \\ 3* \\ 5* \\ 7* \\ 9 \\ 11* \\ 13* \\ 15 \\ 17* \\ 19* \\ 21 \end{pmatrix}, A_{21}{}'' = \begin{pmatrix} 41* \\ 39 \\ 37* \\ 35 \\ 33 \\ 31* \\ 29* \\ 27 \\ 25 \\ 23* \\ 21 \end{pmatrix} \rightarrow$$

$$\rightarrow A_{21}* = A_{21}{}' \cap * A_{21}{}'' = \begin{pmatrix} 5* & 37* \\ 11* & 31* \\ 19* & 23* \end{pmatrix}.$$

$2n=44$

$$A_{22}=\begin{pmatrix} 1 & 43* \\ 3* & 41* \\ 5* & 39 \\ 7* & 37* \\ 9 & 35 \\ 11* & 33 \\ 13* & 31* \\ 15 & 29* \\ 17* & 27 \\ 19* & 25 \\ 21 & 23* \end{pmatrix}, A_{22}{}'=\begin{pmatrix} 1 \\ 3* \\ 5* \\ 7* \\ 9 \\ 11* \\ 13* \\ 15 \\ 17* \\ 19* \\ 21 \end{pmatrix}, A_{22}{}''=\begin{pmatrix} 43* \\ 41* \\ 39 \\ 37* \\ 35 \\ 33 \\ 31* \\ 29* \\ 27 \\ 25 \\ 23* \end{pmatrix} \rightarrow A_{22}*=A_{22}{}' \bigcap * A_{22}{}''=\begin{pmatrix} 3* & 41* \\ 7* & 37* \\ 13* & 31* \end{pmatrix}$$

$2n=46$

$$A_{23}=\begin{pmatrix} 1 & 45 \\ 3* & 43* \\ 5* & 41* \\ 7* & 39 \\ 9 & 37* \\ 11* & 35 \\ 13* & 33 \\ 15 & 31* \\ 17* & 29* \\ 19* & 27 \\ 21 & 25 \\ 23* & 23* \end{pmatrix}, A_{23}{}'=\begin{pmatrix} 1 \\ 3* \\ 5* \\ 7* \\ 9 \\ 11* \\ 13* \\ 15 \\ 17* \\ 19* \\ 21 \\ 23* \end{pmatrix}, A_{23}{}''=\begin{pmatrix} 45 \\ 43* \\ 41* \\ 39 \\ 37* \\ 35 \\ 33 \\ 31* \\ 29* \\ 27 \\ 25 \\ 23* \end{pmatrix} \rightarrow$$

$$\rightarrow A_{23}*=A_{23}{}' \bigcap * A_{23}{}''=\begin{pmatrix} 3* & 43* \\ 5* & 41* \\ 17* & 29* \\ 23* & 23* \end{pmatrix}$$

$2n=48$

$$A_{24} = \begin{pmatrix} 1 & 47* \\ 3* & 45 \\ 5* & 43* \\ 7* & 41* \\ 9 & 39 \\ 11* & 37* \\ 13* & 35 \\ 15 & 33 \\ 17* & 31* \\ 19* & 29* \\ 21 & 27 \\ 23* & 25 \end{pmatrix}, A_{24}{}^{/} = \begin{pmatrix} 1 \\ 3* \\ 5* \\ 7* \\ 9 \\ 11* \\ 13* \\ 15 \\ 17* \\ 19* \\ 21 \\ 23* \end{pmatrix}, A_{24}{}^{//} = \begin{pmatrix} 47* \\ 45 \\ 43* \\ 41* \\ 39 \\ 37* \\ 35 \\ 33 \\ 31* \\ 29* \\ 27 \\ 25 \end{pmatrix} \rightarrow A_{24}* = A_{24}{}^{/} \bigcap * A_{24}{}^{//} = \begin{pmatrix} 5* & 43* \\ 7* & 41* \\ 11* & 37* \\ 17* & 31* \\ 19* & 29* \end{pmatrix}$$

$2n=50$

$$A_{25} = \begin{pmatrix} 1 & 49 \\ 3* & 47* \\ 5* & 45 \\ 7* & 43* \\ 9 & 41* \\ 11* & 39 \\ 13* & 37* \\ 15 & 35 \\ 17* & 33 \\ 19* & 31* \\ 21 & 29* \\ 23* & 27 \\ 25 & 25 \end{pmatrix}, A_{25}{}^{/} = \begin{pmatrix} 1 \\ 3* \\ 5* \\ 7* \\ 9 \\ 11* \\ 13* \\ 15 \\ 17* \\ 19* \\ 21 \\ 23* \\ 25 \end{pmatrix}, A_{25}{}^{//} = \begin{pmatrix} 49 \\ 47* \\ 45 \\ 43* \\ 41* \\ 39 \\ 37* \\ 35 \\ 33 \\ 31* \\ 29* \\ 27 \\ 25 \end{pmatrix} \rightarrow$$

$$\rightarrow A_{25}* = A_{25}{}^{/} \bigcap * A_{25}{}^{//} = \begin{pmatrix} 3* & 47* \\ 7* & 43* \\ 13* & 37* \\ 19* & 31* \end{pmatrix}$$

$2n=52$

$$A_{26} = \begin{pmatrix} 1 & 51 \\ 3* & 49 \\ 5* & 47* \\ 7* & 45 \\ 9 & 43* \\ 11* & 41* \\ 13* & 39 \\ 15 & 37 \\ 17* & 35 \\ 19* & 33 \\ 21 & 31* \\ 23* & 29* \\ 25 & 27 \end{pmatrix}, A_{26}' = \begin{pmatrix} 1 \\ 3* \\ 5* \\ 7* \\ 9 \\ 11* \\ 13* \\ 15 \\ 17* \\ 19* \\ 21 \\ 23* \\ 25 \end{pmatrix}, A_{26}'' = \begin{pmatrix} 51 \\ 49 \\ 47* \\ 45 \\ 43* \\ 41* \\ 39 \\ 37* \\ 35 \\ 33 \\ 31* \\ 29* \\ 27 \end{pmatrix} \rightarrow A_{26}* = A_{26}' \bigcap * A_{26}'' = \begin{pmatrix} 5* & 47* \\ 11* & 41* \\ 23* & 29* \end{pmatrix}$$

$2n=54$

$$A_{27} = \begin{pmatrix} 1 & 53* \\ 3* & 51 \\ 5* & 49 \\ 7* & 47* \\ 9 & 45 \\ 11* & 43* \\ 13* & 41* \\ 15 & 39 \\ 17* & 37* \\ 19* & 35 \\ 21 & 33 \\ 23* & 31* \\ 25 & 29* \\ 27 & 27 \end{pmatrix}, A_{27}' = \begin{pmatrix} 1 \\ 3* \\ 5* \\ 7* \\ 9 \\ 11* \\ 13* \\ 15 \\ 17* \\ 19* \\ 21 \\ 23* \\ 25 \\ 27 \end{pmatrix}, A_{27}'' = \begin{pmatrix} 53* \\ 51 \\ 49 \\ 47* \\ 45 \\ 43* \\ 41* \\ 39 \\ 37* \\ 35 \\ 33 \\ 31* \\ 29* \\ 27 \end{pmatrix} \rightarrow A_{27}* = A_{27}' \bigcap * A_{27}'' = \begin{pmatrix} 7* & 47* \\ 11* & 43* \\ 13* & 41* \\ 17* & 37* \\ 23* & 31* \end{pmatrix}$$

$2n = 56$

$$A_{28} = \begin{pmatrix} 1 & 55 \\ 3* & 53* \\ 5* & 51 \\ 7* & 49 \\ 9 & 47* \\ 11* & 45 \\ 13* & 43* \\ 15 & 41* \\ 17* & 39 \\ 19* & 37* \\ 21 & 35 \\ 23* & 33 \\ 25 & 31* \\ 27 & 29* \end{pmatrix} A_{28}{}' = \begin{pmatrix} 1 \\ 3* \\ 5* \\ 7* \\ 9 \\ 11* \\ 13* \\ 15 \\ 17* \\ 19* \\ 21 \\ 23* \\ 25 \\ 27 \end{pmatrix} A_{28}{}'' = \begin{pmatrix} 55 \\ 53* \\ 51 \\ 49 \\ 47* \\ 45 \\ 43* \\ 41* \\ 39 \\ 37* \\ 35 \\ 33 \\ 31* \\ 29* \end{pmatrix} \rightarrow A_{28}* = A_{28}{}' \cap * A_{28}{}'' = \begin{pmatrix} 3* & 53* \\ 13* & 43* \\ 19* & 37* \end{pmatrix}$$

$n = 58$

$$A_{29} = \begin{pmatrix} 1 & 57 \\ 3* & 55 \\ 5* & 53* \\ 7* & 51 \\ 9 & 49 \\ 11* & 47* \\ 13* & 45 \\ 15 & 43* \\ 17* & 41* \\ 19* & 39 \\ 21 & 37* \\ 23* & 35 \\ 25 & 33 \\ 27 & 31* \\ 29* & 29* \end{pmatrix} A_{29}{}' = \begin{pmatrix} 1 \\ 3* \\ 5* \\ 7* \\ 9 \\ 11* \\ 13* \\ 15 \\ 17* \\ 19* \\ 21 \\ 23* \\ 25 \\ 27 \\ 29* \end{pmatrix} A_{29}{}'' = \begin{pmatrix} 57 \\ 55 \\ 53* \\ 51 \\ 49 \\ 47* \\ 45 \\ 43* \\ 41* \\ 39 \\ 37* \\ 35 \\ 33 \\ 31* \\ 29* \end{pmatrix}, A_{29}* = A_{29}{}' \cap * A_{29}{}'' = \begin{pmatrix} 5* & 53* \\ 11* & 47* \\ 17* & 41* \\ 29* & 29* \end{pmatrix}$$

$2n=60$

$$A_{30} = \begin{pmatrix} 1 & 59* \\ 3* & 57 \\ 5* & 55 \\ 7* & 53* \\ 9 & 51 \\ 11* & 49 \\ 13* & 47* \\ 15 & 45 \\ 17* & 43* \\ 19* & 41* \\ 21 & 39 \\ 23* & 37* \\ 25 & 35 \\ 27 & 33 \\ 29* & 31* \end{pmatrix}, A_{30}{}' = \begin{pmatrix} 1 \\ 3* \\ 5* \\ 7* \\ 9 \\ 11* \\ 13* \\ 15 \\ 17* \\ 19* \\ 21 \\ 23* \\ 25 \\ 27 \\ 29* \end{pmatrix}, A_{30}{}'' = \begin{pmatrix} 59* \\ 57 \\ 55 \\ 53* \\ 51 \\ 49 \\ 47* \\ 45 \\ 43* \\ 41* \\ 39 \\ 37* \\ 35 \\ 33 \\ 31* \end{pmatrix}, \rightarrow A_{30}{}* = A_{30}{}' \cap *A_{30}{}'' = \begin{pmatrix} 7* & 53* \\ 13* & 47* \\ 17* & 43* \\ 19* & 41* \\ 23* & 37* \\ 29* & 31* \end{pmatrix}$$

$2n=62$

$$A_{31} = \begin{pmatrix} 1 & 61* \\ 3* & 59* \\ 5* & 57 \\ 7* & 55 \\ 9 & 53* \\ 11* & 51 \\ 13* & 49 \\ 15 & 47* \\ 17* & 45 \\ 19* & 43* \\ 21 & 41* \\ 23* & 39 \\ 25 & 37* \\ 27 & 35 \\ 29* & 33 \\ 31* & 31* \end{pmatrix}, A_{31}{}' = \begin{pmatrix} 1 \\ 3* \\ 5* \\ 7* \\ 9 \\ 11* \\ 13* \\ 15 \\ 17* \\ 19* \\ 21 \\ 23* \\ 25 \\ 27 \\ 29* \\ 31* \end{pmatrix}, A_{31}{}'' = \begin{pmatrix} 61* \\ 59* \\ 57 \\ 55 \\ 53* \\ 51 \\ 49 \\ 47* \\ 45 \\ 43* \\ 41* \\ 39 \\ 37* \\ 35 \\ 33 \\ 31* \end{pmatrix}, \rightarrow A_{31}{}* = A_{31}{}' \cap *A_{31}{}'' = \begin{pmatrix} 3* & 59* \\ 19* & 43* \\ 31* & 31* \end{pmatrix}$$

$2n=64$

$$A_{32} = \begin{pmatrix} 1 & 63 \\ 3* & 61* \\ 5* & 59* \\ 7* & 57 \\ 9 & 55 \\ 11* & 53* \\ 13* & 51 \\ 15 & 49 \\ 17* & 47* \\ 19* & 45 \\ 21 & 43* \\ 23* & 41* \\ 25 & 39 \\ 27 & 37* \\ 29* & 35 \\ 31* & 33 \end{pmatrix}, A_{32}' = \begin{pmatrix} 1 \\ 3* \\ 5* \\ 7* \\ 9 \\ 11* \\ 13* \\ 15 \\ 17* \\ 19* \\ 21 \\ 23* \\ 25 \\ 27 \\ 29* \\ 31* \end{pmatrix}, A_{32}'' = \begin{pmatrix} 63 \\ 61* \\ 59* \\ 57 \\ 55 \\ 53* \\ 51 \\ 49 \\ 47* \\ 45 \\ 43* \\ 41* \\ 39 \\ 37* \\ 35 \\ 33 \end{pmatrix}, \rightarrow A_{32}* = A_{32}' \bigcap *A_{32}'' = \begin{pmatrix} 3* & 61* \\ 5* & 59* \\ 11* & 53* \\ 17* & 47* \\ 23* & 41* \end{pmatrix}$$

$2n=66$

$$A_{33} = \begin{pmatrix} 1 & 65 \\ 3* & 63 \\ 5* & 61* \\ 7* & 59* \\ 9 & 57 \\ 11* & 55 \\ 13* & 53* \\ 15 & 51 \\ 17* & 49 \\ 19* & 47* \\ 21 & 45 \\ 23* & 43* \\ 25 & 41* \\ 27 & 39 \\ 29* & 37* \\ 31* & 35 \\ 33 & 33 \end{pmatrix}, A_{33}' = \begin{pmatrix} 1 \\ 3* \\ 5* \\ 7* \\ 9 \\ 11* \\ 13* \\ 15 \\ 17* \\ 19* \\ 21 \\ 23* \\ 25 \\ 27 \\ 29* \\ 31* \\ 33 \end{pmatrix}, A_{33}'' = \begin{pmatrix} 65 \\ 63 \\ 61* \\ 59* \\ 57 \\ 55 \\ 53* \\ 51 \\ 49 \\ 47* \\ 45 \\ 43* \\ 41* \\ 39 \\ 37* \\ 35 \\ 33 \end{pmatrix}, \rightarrow A_{33}* = A_{33}' \bigcap *A_{33}'' = \begin{pmatrix} 5* & 61* \\ 7* & 59* \\ 13* & 53* \\ 19* & 47* \\ 23* & 43* \\ 29* & 37* \end{pmatrix}$$

$2n=68$

$$A_{34} = \begin{pmatrix} 1 & 67* \\ 3* & 65 \\ 5* & 63 \\ 7* & 61* \\ 9 & 59* \\ 11* & 57 \\ 13* & 55 \\ 15 & 53* \\ 17* & 51 \\ 19* & 49 \\ 21 & 47* \\ 23* & 45 \\ 25 & 43* \\ 27 & 41* \\ 29* & 39 \\ 31* & 37* \\ 33 & 35 \end{pmatrix}, A_{34}{}' = \begin{pmatrix} 1 \\ 3* \\ 5* \\ 7* \\ 9 \\ 11* \\ 13* \\ 15 \\ 17* \\ 19* \\ 21 \\ 23* \\ 25 \\ 27 \\ 29* \\ 31* \\ 33 \end{pmatrix}, A_{34}{}'' = \begin{pmatrix} 67* \\ 65 \\ 63 \\ 61* \\ 59* \\ 57 \\ 55 \\ 53* \\ 51 \\ 49 \\ 47* \\ 45 \\ 43* \\ 41* \\ 39 \\ 37* \\ 35 \end{pmatrix}, \rightarrow$$

$$\rightarrow A_{34}* = A_{34}{}' \cap *A_{34}{}'' = \begin{pmatrix} 7* & 61* \\ 31* & 37* \end{pmatrix}$$

$$2n=70$$

$$A_{35} = \begin{pmatrix} 1 & 69 \\ 3* & 67* \\ 5* & 65 \\ 7* & 63 \\ 9 & 61* \\ 11* & 59* \\ 13* & 57 \\ 15 & 55 \\ 17* & 53* \\ 19* & 51 \\ 21 & 49 \\ 23* & 47* \\ 25 & 45 \\ 27 & 43* \\ 29* & 41* \\ 31* & 39 \\ 33 & 37* \\ 35 & 35 \end{pmatrix}, A_{35}{}' = \begin{pmatrix} 1 \\ 3* \\ 5* \\ 7* \\ 9 \\ 11* \\ 13* \\ 15 \\ 17* \\ 19* \\ 21 \\ 23* \\ 25 \\ 27 \\ 29* \\ 31* \\ 33 \\ 35 \end{pmatrix}, A_{35}{}'' = \begin{pmatrix} 69 \\ 67* \\ 65 \\ 63 \\ 61* \\ 59* \\ 57 \\ 55 \\ 53* \\ 51 \\ 49 \\ 47* \\ 45 \\ 43* \\ 41* \\ 39 \\ 37* \\ 35 \end{pmatrix}, \rightarrow$$

$$\rightarrow A_{35}* = A_{35}{}' \bigcap * A_{35}{}'' = \begin{pmatrix} 3* & 67* \\ 11* & 59* \\ 17* & 53* \\ 23* & 47* \\ 29* & 41* \end{pmatrix}$$

$2n=72$

$$A_{36} = \begin{pmatrix} 1 & 71 \\ 3* & 69 \\ 5* & 67* \\ 7* & 65 \\ 9 & 63 \\ 11* & 61* \\ 13* & 59* \\ 15 & 57 \\ 17* & 55 \\ 19* & 53* \\ 21 & 51 \\ 23* & 49 \\ 25 & 47* \\ 27 & 45 \\ 29* & 43* \\ 31* & 41* \\ 33 & 39 \\ 35 & 37* \end{pmatrix}, A_{36}{}' = \begin{pmatrix} 1 \\ 3* \\ 5* \\ 7* \\ 9 \\ 11* \\ 13* \\ 15 \\ 17* \\ 19* \\ 21 \\ 23* \\ 25 \\ 27 \\ 29* \\ 31* \\ 33 \\ 35 \end{pmatrix}, A_{36}{}'' = \begin{pmatrix} 71 \\ 69 \\ 67* \\ 65 \\ 63 \\ 61* \\ 59* \\ 57 \\ 55 \\ 53* \\ 51 \\ 49 \\ 47* \\ 45 \\ 43* \\ 41* \\ 39 \\ 37* \end{pmatrix}, \rightarrow$$

$$\rightarrow A_{36}* = A_{36}{}' \bigcap *A_{36}{}'' = \begin{pmatrix} 5* & 67* \\ 11* & 61* \\ 13* & 59* \\ 19* & 53* \\ 29* & 43* \\ 31* & 41* \end{pmatrix}$$

26

$2n=74$

$$A_{37} = \begin{pmatrix} 1 & 73 \\ 3* & 71 \\ 5* & 69 \\ 7* & 67* \\ 9 & 65 \\ 11* & 63 \\ 13* & 61* \\ 15 & 59* \\ 17* & 57 \\ 19* & 55 \\ 21 & 53* \\ 23* & 51 \\ 25 & 49 \\ 27 & 47* \\ 29* & 45 \\ 31* & 43* \\ 33 & 41* \\ 35 & 39 \\ 37* & 37* \end{pmatrix}, A_{37}{}' = \begin{pmatrix} 1 \\ 3* \\ 5* \\ 7* \\ 9 \\ 11* \\ 13* \\ 15 \\ 17* \\ 19* \\ 21 \\ 23* \\ 25 \\ 27 \\ 29* \\ 31* \\ 33 \\ 35 \\ 37* \end{pmatrix}, A_{37}{}'' = \begin{pmatrix} 73 \\ 71 \\ 69 \\ 67* \\ 65 \\ 63 \\ 61* \\ 59* \\ 57 \\ 55 \\ 53* \\ 51 \\ 49 \\ 47* \\ 45 \\ 43* \\ 41* \\ 39 \\ 37* \end{pmatrix}, \rightarrow$$

$$\rightarrow A_{37}* = A_{37}{}' \bigcap *A_{37}{}'' = \begin{pmatrix} 7* & 67* \\ 13* & 61* \\ 31* & 43* \\ 37* & 37* \end{pmatrix}$$

$$2n=76$$

$$A_{38} = \begin{pmatrix} 1 & 75 \\ 3* & 73* \\ 5* & 71* \\ 7* & 69 \\ 9 & 67* \\ 11* & 65 \\ 13* & 63 \\ 15 & 61* \\ 17* & 59* \\ 19* & 57 \\ 21 & 55 \\ 23* & 53* \\ 25 & 51 \\ 27 & 49 \\ 29* & 47* \\ 31* & 45 \\ 33 & 43* \\ 35 & 41* \\ 37* & 39 \end{pmatrix}, A_{38}^{/} = \begin{pmatrix} 1 \\ 3* \\ 5* \\ 7* \\ 9 \\ 11* \\ 13* \\ 15 \\ 17* \\ 19* \\ 21 \\ 23* \\ 25 \\ 27 \\ 29* \\ 31* \\ 33 \\ 35 \\ 37* \end{pmatrix}, A_{38}^{//} = \begin{pmatrix} 75 \\ 73* \\ 71* \\ 69 \\ 67* \\ 65 \\ 63 \\ 61* \\ 59* \\ 57 \\ 55 \\ 53* \\ 51 \\ 49 \\ 47* \\ 45 \\ 43* \\ 41* \\ 39 \end{pmatrix}, \rightarrow$$

$$\rightarrow A_{38}* = A_{38}^{/} \cap *A_{38}^{//} = \begin{pmatrix} 3* & 73* \\ 5* & 71* \\ 17* & 59* \\ 23* & 53* \\ 29* & 47* \end{pmatrix}$$

28

$2n=78$

$$A_{39} = \begin{pmatrix} 1 & 77 \\ 3* & 75 \\ 5* & 73* \\ 7* & 71* \\ 9 & 69 \\ 11* & 67* \\ 13* & 65 \\ 15 & 63 \\ 17* & 61* \\ 19* & 59* \\ 21 & 57 \\ 23* & 55 \\ 25 & 53* \\ 27 & 51 \\ 29* & 49 \\ 31* & 47* \\ 33 & 45 \\ 35 & 43* \\ 37* & 41* \\ 39 & 39 \end{pmatrix}, A_{39}{}' = \begin{pmatrix} 1 \\ 3* \\ 5* \\ 7* \\ 9 \\ 11* \\ 13* \\ 15 \\ 17* \\ 19* \\ 21 \\ 23* \\ 25 \\ 27 \\ 29* \\ 31* \\ 33 \\ 35 \\ 37* \\ 39 \end{pmatrix}, A_{39}{}'' = \begin{pmatrix} 77 \\ 75 \\ 73* \\ 71* \\ 69 \\ 67* \\ 65 \\ 63 \\ 61* \\ 59* \\ 57 \\ 55 \\ 53* \\ 51 \\ 49 \\ 47* \\ 45 \\ 43* \\ 41* \\ 39 \end{pmatrix}, \rightarrow$$

$$\rightarrow A_{39}* = A_{39}{}' \bigcap *A_{39}{}'' = \begin{pmatrix} 5* & 73* \\ 7* & 71* \\ 11* & 67* \\ 17* & 61* \\ 19* & 59* \\ 31* & 47* \\ 37* & 41* \end{pmatrix}$$

$2n=80$

$$A_{40} = \begin{pmatrix} 1 & 79* \\ 3* & 77 \\ 5* & 75 \\ 7* & 73* \\ 9 & 71* \\ 11* & 69 \\ 13* & 67* \\ 15 & 65 \\ 17* & 63 \\ 19* & 61* \\ 21 & 59* \\ 23* & 57 \\ 25 & 55 \\ 27 & 53* \\ 29* & 51 \\ 31* & 49 \\ 33 & 47* \\ 35 & 45 \\ 37* & 43* \\ 39 & 41* \end{pmatrix}, A_{40}{}' = \begin{pmatrix} 1 \\ 3* \\ 5* \\ 7* \\ 9 \\ 11* \\ 13* \\ 15 \\ 17* \\ 19* \\ 21 \\ 23* \\ 25 \\ 27 \\ 29* \\ 31* \\ 33 \\ 35 \\ 37* \\ 39 \end{pmatrix}, A_{40}{}'' = \begin{pmatrix} 79* \\ 77 \\ 75 \\ 73* \\ 71* \\ 69 \\ 67* \\ 65 \\ 63 \\ 61* \\ 59* \\ 57 \\ 55 \\ 53* \\ 51 \\ 49 \\ 47* \\ 45 \\ 43* \\ 41* \end{pmatrix}, \rightarrow$$

$$\rightarrow A_{40}* = A_{40}{}' \cap *A_{40}{}'' = \begin{pmatrix} 7* & 73* \\ 13* & 67* \\ 19* & 61* \\ 37* & 43* \end{pmatrix}$$

$2n=82$

$$A_{41} = \begin{pmatrix} 1 & 81 \\ 3* & 79* \\ 5* & 77 \\ 7* & 75 \\ 9 & 73* \\ 11* & 71* \\ 13* & 69 \\ 15 & 67* \\ 17* & 65 \\ 19* & 63 \\ 21 & 61* \\ 23* & 59* \\ 25 & 57 \\ 27 & 55 \\ 29* & 53* \\ 31* & 51 \\ 33 & 49 \\ 35 & 47* \\ 37* & 45 \\ 39 & 43* \\ 41* & 41* \end{pmatrix}, A_{41}{}' = \begin{pmatrix} 1 \\ 3* \\ 5* \\ 7* \\ 9 \\ 11* \\ 13* \\ 15 \\ 17* \\ 19* \\ 21 \\ 23* \\ 25 \\ 27 \\ 29* \\ 31* \\ 33 \\ 35 \\ 37* \\ 39 \\ 41* \end{pmatrix}, A_{41}{}'' = \begin{pmatrix} 81 \\ 79* \\ 77 \\ 75 \\ 73* \\ 71* \\ 69 \\ 67* \\ 65 \\ 63 \\ 61* \\ 59* \\ 57 \\ 55 \\ 53* \\ 51 \\ 49 \\ 47* \\ 45 \\ 43* \\ 41* \end{pmatrix}, \rightarrow$$

$$\rightarrow A_{41}* = A_{41}{}' \cap *A_{41}{}'' = \begin{pmatrix} 3* & 79* \\ 11* & 71* \\ 23* & 59* \\ 29* & 53* \\ 41* & 41* \end{pmatrix}$$

$2n=84$

$$A_{42} = \begin{pmatrix} 1 & 83* \\ 3* & 81 \\ 5* & 79* \\ 7* & 77 \\ 9 & 75 \\ 11* & 73* \\ 13* & 71* \\ 15 & 69 \\ 17* & 67* \\ 19* & 65 \\ 21 & 63 \\ 23* & 61* \\ 25 & 59* \\ 27 & 57 \\ 29* & 55 \\ 31* & 53* \\ 33 & 51 \\ 35 & 49 \\ 37* & 47* \\ 39 & 45 \\ 41* & 43* \end{pmatrix}, A_{42}{}' = \begin{pmatrix} 1 \\ 3* \\ 5* \\ 7* \\ 9 \\ 11* \\ 13* \\ 15 \\ 17* \\ 19* \\ 21 \\ 23* \\ 25 \\ 27 \\ 29* \\ 31* \\ 33 \\ 35 \\ 37* \\ 39 \\ 41* \end{pmatrix}, A_{42}{}'' = \begin{pmatrix} 83* \\ 81 \\ 79* \\ 77 \\ 75 \\ 73* \\ 71* \\ 69 \\ 67* \\ 65 \\ 63 \\ 61* \\ 59* \\ 57 \\ 55 \\ 53* \\ 51 \\ 49 \\ 47* \\ 45 \\ 43* \end{pmatrix}, \rightarrow$$

$$\rightarrow A_{42}* = A_{42}{}' \cap *A_{42}{}'' = \begin{pmatrix} 5* & 79* \\ 11* & 73* \\ 13* & 71* \\ 17* & 67* \\ 23* & 61* \\ 31* & 53* \\ 37* & 47* \\ 41* & 43* \end{pmatrix}$$

$2n=86$

$$A_{43} = \begin{pmatrix} 1 & 85 \\ 3* & 83* \\ 5* & 81 \\ 7* & 79* \\ 9 & 77 \\ 11* & 75 \\ 13* & 73* \\ 15 & 71* \\ 17* & 69 \\ 19* & 67* \\ 21 & 65 \\ 23* & 63 \\ 25 & 61* \\ 27 & 59* \\ 29* & 57 \\ 31* & 55 \\ 33 & 53* \\ 35 & 51 \\ 37* & 49 \\ 39 & 47* \\ 41* & 45 \\ 43* & 43* \end{pmatrix}, \; A_{43}{}' = \begin{pmatrix} 1 \\ 3* \\ 5* \\ 7* \\ 9 \\ 11* \\ 13* \\ 15 \\ 17* \\ 19* \\ 21 \\ 23* \\ 25 \\ 27 \\ 29* \\ 31* \\ 33 \\ 35 \\ 37* \\ 39 \\ 41* \\ 43* \end{pmatrix}, A_{43}{}'' = \begin{pmatrix} 85 \\ 83* \\ 81 \\ 79* \\ 77 \\ 75 \\ 73* \\ 71* \\ 69 \\ 67* \\ 65 \\ 63 \\ 61* \\ 59* \\ 57 \\ 55 \\ 53* \\ 51 \\ 49 \\ 47* \\ 45 \\ 43* \end{pmatrix} \rightarrow$$

$$\rightarrow A_{43}* = A_{43}{}' \cap * A_{43}{}'' = \begin{pmatrix} 3*83* \\ 7*79* \\ 13*73* \\ 19*67* \\ 43*43* \end{pmatrix}$$

$$2n=88$$

$$A_{44} = \begin{pmatrix} 1 & 87 \\ 3* & 85 \\ 5* & 83* \\ 7* & 81 \\ 9 & 79* \\ 11* & 77 \\ 13* & 75 \\ 15 & 73* \\ 17* & 71* \\ 19* & 69 \\ 21 & 67* \\ 23* & 65 \\ 25 & 63 \\ 27 & 61* \\ 29* & 59* \\ 31* & 57 \\ 33 & 55 \\ 35 & 53* \\ 37* & 51 \\ 39 & 49 \\ 41* & 47* \\ 43* & 45 \end{pmatrix}, A_{44}{}' = \begin{pmatrix} 1 \\ 3* \\ 5* \\ 7* \\ 9 \\ 11* \\ 13* \\ 15 \\ 17* \\ 19* \\ 21 \\ 23* \\ 25 \\ 27 \\ 29* \\ 31* \\ 33 \\ 35 \\ 37* \\ 39 \\ 41* \\ 43* \end{pmatrix}, A_{44}{}'' = \begin{pmatrix} 87 \\ 85 \\ 83* \\ 81 \\ 79* \\ 77 \\ 75 \\ 73* \\ 71* \\ 69 \\ 67* \\ 65 \\ 63 \\ 61* \\ 59* \\ 57 \\ 55 \\ 53* \\ 51 \\ 49 \\ 47* \\ 45 \end{pmatrix}, \rightarrow$$

$$\rightarrow A_{44}* = A_{44}{}' \bigcap *A_{44}{}'' = \begin{pmatrix} 5* & 83* \\ 17* & 71* \\ 29* & 59* \\ 41* & 47* \end{pmatrix}$$

$2n=90$

$$A_{45} = \begin{pmatrix} 1 & 89* \\ 3* & 87 \\ 5* & 85 \\ 7* & 83* \\ 9 & 81 \\ 11* & 79* \\ 13* & 77 \\ 15 & 75 \\ 17* & 73* \\ 19* & 71* \\ 21 & 69 \\ 23* & 67* \\ 25 & 65 \\ 27 & 63 \\ 29* & 61* \\ 31* & 59* \\ 33 & 57 \\ 35 & 55 \\ 37* & 53* \\ 39 & 51 \\ 41* & 49 \\ 43* & 47* \\ 45 & 45 \end{pmatrix}, A_{45}^{\prime} = \begin{pmatrix} 1 \\ 3* \\ 5* \\ 7* \\ 9 \\ 11* \\ 13* \\ 15 \\ 17* \\ 19* \\ 21 \\ 23* \\ 25 \\ 27 \\ 29* \\ 31* \\ 33 \\ 35 \\ 37* \\ 39 \\ 41* \\ 43* \\ 45 \end{pmatrix}, A_{45}^{\prime\prime} = \begin{pmatrix} 89 \\ 87 \\ 85 \\ 83* \\ 81 \\ 79* \\ 77 \\ 75 \\ 73* \\ 71* \\ 69 \\ 67* \\ 65 \\ 63 \\ 61* \\ 59* \\ 57 \\ 55 \\ 53* \\ 51 \\ 49 \\ 47* \\ 45 \end{pmatrix}, \rightarrow$$

$$\rightarrow A_{45}^* = A_{45}^{\prime} \bigcap {}^* A_{45}^{\prime\prime} = \begin{pmatrix} 7* & 83* \\ 11* & 79* \\ 17* & 73* \\ 19* & 71* \\ 23* & 67* \\ 29* & 61* \\ 31* & 59* \\ 37* & 53* \\ 43* & 47* \end{pmatrix}$$

$$2n=92$$

$$A_{46} = \begin{pmatrix} 1 & 91 \\ 3* & 89* \\ 5* & 87 \\ 7* & 85 \\ 9 & 83* \\ 11* & 81 \\ 13* & 79* \\ 15 & 77 \\ 17* & 75 \\ 19* & 73* \\ 21 & 71* \\ 23* & 69 \\ 25 & 67* \\ 27 & 65 \\ 29* & 63 \\ 31* & 61* \\ 33 & 59* \\ 35 & 57 \\ 37* & 55 \\ 39 & 53* \\ 41* & 51 \\ 43* & 49 \\ 45 & 47* \end{pmatrix}, A_{46}{}' = \begin{pmatrix} 1 \\ 3* \\ 5* \\ 7* \\ 9 \\ 11* \\ 13* \\ 15 \\ 17* \\ 19* \\ 21 \\ 23* \\ 25 \\ 27 \\ 29* \\ 31* \\ 33 \\ 35 \\ 37* \\ 39 \\ 41* \\ 43* \\ 45 \end{pmatrix}, A_{46}{}'' = \begin{pmatrix} 91 \\ 89* \\ 87 \\ 85 \\ 83* \\ 81 \\ 79* \\ 77 \\ 75 \\ 73* \\ 71* \\ 69 \\ 67* \\ 65 \\ 63 \\ 61* \\ 59* \\ 57 \\ 55 \\ 53* \\ 51 \\ 49 \\ 47* \end{pmatrix}, \rightarrow$$

$$\rightarrow A_{46}* = A_{46}{}' \cap * A_{46}{}'' = \begin{pmatrix} 3* & 89* \\ 13* & 79* \\ 19* & 73* \\ 31* & 61* \end{pmatrix}$$

$2n=94$

$$A_{47} = \begin{pmatrix} 1 & 93 \\ 3* & 91 \\ 5* & 89* \\ 7* & 87 \\ 9 & 85 \\ 11* & 83* \\ 13* & 81 \\ 15 & 79* \\ 17* & 77 \\ 19* & 75 \\ 21 & 73* \\ 23* & 71* \\ 25 & 69 \\ 27 & 67* \\ 29* & 65 \\ 31* & 63 \\ 33 & 61* \\ 35 & 59* \\ 37* & 57 \\ 39 & 55 \\ 41* & 53* \\ 43* & 51 \\ 45 & 49 \\ 47* & 47* \end{pmatrix}, A_{47}{}' = \begin{pmatrix} 1 \\ 3* \\ 5* \\ 7* \\ 9 \\ 11* \\ 13* \\ 15 \\ 17* \\ 19* \\ 21 \\ 23* \\ 25 \\ 27 \\ 29* \\ 31* \\ 33 \\ 35 \\ 37* \\ 39 \\ 41* \\ 43* \\ 45 \\ 47* \end{pmatrix}, A_{47}{}'' = \begin{pmatrix} 93 \\ 91 \\ 89* \\ 87 \\ 85 \\ 83* \\ 81 \\ 79* \\ 77 \\ 75 \\ 73* \\ 71* \\ 69 \\ 67* \\ 65 \\ 63 \\ 61* \\ 59* \\ 57 \\ 55 \\ 53* \\ 51 \\ 49 \\ 47* \end{pmatrix}, \rightarrow$$

$$\rightarrow A_{47}* = A_{47}{}' \bigcap *A_{47}{}'' = \begin{pmatrix} 5* & 89* \\ 11* & 83* \\ 23* & 71* \\ 41* & 53* \\ 47* & 47* \end{pmatrix}$$

$$2n=96$$

$$A_{48} = \begin{pmatrix} 1 & 95 \\ 3* & 93 \\ 5* & 91 \\ 7* & 89* \\ 9 & 87 \\ 11* & 85 \\ 13* & 83* \\ 15 & 81 \\ 17* & 79* \\ 19* & 77 \\ 21 & 75 \\ 23* & 73* \\ 25 & 71* \\ 27 & 69 \\ 29* & 67* \\ 31* & 65 \\ 33 & 63 \\ 35 & 61* \\ 37* & 59* \\ 39 & 57 \\ 41* & 55 \\ 43* & 53* \\ 45 & 51 \\ 47* & 49 \end{pmatrix}, A_{48}{}' = \begin{pmatrix} 1 \\ 3* \\ 5* \\ 7* \\ 9 \\ 11* \\ 13* \\ 15 \\ 17* \\ 19* \\ 21 \\ 23* \\ 25 \\ 27 \\ 29* \\ 31* \\ 33 \\ 35 \\ 37* \\ 39 \\ 41* \\ 43* \\ 45 \\ 47* \end{pmatrix}, A_{48}{}'' = \begin{pmatrix} 95 \\ 93 \\ 91 \\ 89* \\ 87 \\ 85 \\ 83* \\ 81 \\ 79* \\ 77 \\ 75 \\ 73* \\ 71* \\ 69 \\ 67* \\ 65 \\ 63 \\ 61* \\ 59* \\ 57 \\ 55 \\ 53* \\ 51 \\ 49 \end{pmatrix}, \rightarrow$$

$$\rightarrow A_{48}{}^* = A_{48}{}' \bigcap {}^*A_{48}{}'' = \begin{pmatrix} 7* & 89* \\ 13* & 83* \\ 17* & 79* \\ 23* & 73* \\ 29* & 67* \\ 37* & 59* \\ 43* & 53* \end{pmatrix}$$

$2n=98$

$$A_{49} = \begin{pmatrix} 1 & 97* \\ 3* & 95 \\ 5* & 93 \\ 7* & 91 \\ 9 & 89* \\ 11* & 87 \\ 13* & 85 \\ 15 & 83* \\ 17* & 81 \\ 19* & 79* \\ 21 & 77 \\ 23* & 75 \\ 25 & 73* \\ 27 & 71* \\ 29* & 69 \\ 31* & 67* \\ 33 & 65 \\ 35 & 63 \\ 37* & 61* \\ 39 & 59* \\ 41* & 57 \\ 43* & 55 \\ 45 & 53* \\ 47* & 51 \\ 49 & 49 \end{pmatrix}, A_{49}{}' = \begin{pmatrix} 1 \\ 3* \\ 5* \\ 7* \\ 9 \\ 11* \\ 13* \\ 15 \\ 17* \\ 19* \\ 21 \\ 23* \\ 25 \\ 27 \\ 29* \\ 31* \\ 33 \\ 35 \\ 37* \\ 39 \\ 41* \\ 43* \\ 45 \\ 47* \\ 49 \end{pmatrix}, A_{49}{}'' = \begin{pmatrix} 97* \\ 95 \\ 93 \\ 91 \\ 89* \\ 87 \\ 85 \\ 83* \\ 81 \\ 79* \\ 77 \\ 75 \\ 73* \\ 71* \\ 69 \\ 67* \\ 65 \\ 63 \\ 61* \\ 59* \\ 57 \\ 55 \\ 53* \\ 51 \\ 49 \end{pmatrix}, \rightarrow$$

$$\rightarrow A_{49}* = A_{49}{}' \bigcap *A_{49}{}'' = \begin{pmatrix} 19* & 79* \\ 31* & 67* \\ 37* & 61* \end{pmatrix}$$

$$2n=100$$

$$A_{50} = \begin{pmatrix} 1 & 99 \\ 3* & 97* \\ 5* & 95 \\ 7* & 93 \\ 9 & 91 \\ 11* & 89* \\ 13* & 87 \\ 15 & 85 \\ 17* & 83* \\ 19* & 81 \\ 21 & 79* \\ 23* & 77 \\ 25 & 75 \\ 27 & 73* \\ 29* & 71* \\ 31* & 69 \\ 33 & 67* \\ 35 & 65 \\ 37* & 63 \\ 39 & 61* \\ 41* & 59* \\ 43* & 57 \\ 45 & 55 \\ 47* & 53* \\ 49 & 51 \end{pmatrix}, A_{50}{}' = \begin{pmatrix} 1 \\ 3* \\ 5* \\ 7* \\ 9 \\ 11* \\ 13* \\ 15 \\ 17* \\ 19* \\ 21 \\ 23* \\ 25 \\ 27 \\ 29* \\ 31* \\ 33 \\ 35 \\ 37* \\ 39 \\ 41* \\ 43* \\ 45 \\ 47* \\ 49 \end{pmatrix}, A_{50}{}'' = \begin{pmatrix} 99 \\ 97* \\ 95 \\ 93 \\ 91 \\ 89* \\ 87 \\ 85 \\ 83* \\ 81 \\ 79* \\ 77 \\ 75 \\ 73* \\ 71* \\ 69 \\ 67* \\ 65 \\ 63 \\ 61* \\ 59* \\ 57 \\ 55 \\ 53* \\ 51 \end{pmatrix}, \rightarrow$$

$$\rightarrow A_{50}* = A_{50}{}' \cap *A_{50}{}'' = \begin{pmatrix} 3* & 97* \\ 11* & 89* \\ 17* & 83* \\ 29* & 71* \\ 41* & 59* \\ 47* & 53* \end{pmatrix}$$

$2n=102$

$$A_{51} = \begin{pmatrix} 1 & 101* \\ 3* & 99 \\ 5* & 97* \\ 7* & 95 \\ 9 & 93 \\ 11* & 91 \\ 13* & 89* \\ 15 & 87 \\ 17* & 85 \\ 19* & 83* \\ 21 & 81 \\ 23* & 79* \\ 25 & 77 \\ 27 & 75 \\ 29* & 73* \\ 31* & 71* \\ 33 & 69 \\ 35 & 67* \\ 37* & 65 \\ 39 & 63 \\ 41* & 61* \\ 43* & 59* \\ 45 & 57 \\ 47* & 55 \\ 49 & 53* \\ 51 & 51 \end{pmatrix}, \quad A_{51}^{/} = \begin{pmatrix} 1 \\ 3* \\ 5* \\ 7* \\ 9 \\ 11* \\ 13* \\ 15 \\ 17* \\ 19* \\ 21 \\ 23* \\ 25 \\ 27 \\ 29* \\ 31* \\ 33 \\ 35 \\ 37* \\ 39 \\ 41* \\ 43* \\ 45 \\ 47* \\ 49 \\ 51 \end{pmatrix}, \quad A_{51}^{//} = \begin{pmatrix} 101* \\ 99 \\ 97* \\ 95 \\ 93 \\ 91 \\ 89* \\ 87 \\ 85 \setminus \\ 83* \\ 81 \\ 79* \\ 77 \\ 75 \\ 73* \\ 71* \\ 69 \\ 67* \\ 65 \\ 63 \\ 61* \\ 59* \\ 57 \\ 55 \\ 53* \\ 51 \end{pmatrix} \rightarrow$$

$$\rightarrow A_{51}* = A_{51}^{/} \bigcap * A_{51}^{//} = \begin{pmatrix} 5* & 97* \\ 13* & 89* \\ 19* & 83* \\ 23* & 79* \\ 29* & 73* \\ 31* & 71* \\ 41* & 61* \\ 43* & 59* \end{pmatrix}$$

$2n=104$

$$A_{52} = \begin{pmatrix} 1 & 103* \\ 3* & 101* \\ 5* & 99 \\ 7* & 97* \\ 9 & 95 \\ 11* & 93 \\ 13* & 91 \\ 15 & 89* \\ 17* & 87 \\ 19* & 85 \\ 21 & 83* \\ 23* & 81 \\ 25 & 79* \\ 27 & 77 \\ 29* & 75 \\ 31* & 73* \\ 33 & 71* \\ 35 & 69 \\ 37* & 67* \\ 39 & 65 \\ 41* & 63 \\ 43* & 61* \\ 45 & 59* \\ 47* & 57 \\ 49 & 55 \\ 51 & 53* \end{pmatrix}, \; A'_{52} = \begin{pmatrix} 1 \\ 3* \\ 5* \\ 7* \\ 9 \\ 11* \\ 13* \\ 15 \\ 17 \\ 19 \\ 21 \\ 23* \\ 25 \\ 27 \\ 29* \\ 31* \\ 33 \\ 35 \\ 37* \\ 39 \\ 41* \\ 43* \\ 45 \\ 47* \\ 49 \\ 51 \end{pmatrix}, \; A''_{52} = \begin{pmatrix} 103* \\ 101* \\ 99 \\ 97* \\ 95 \\ 93 \\ 91 \\ 89* \\ 87 \\ 85 \\ 83* \\ 81 \\ 79* \\ 77 \\ 75 \\ 73* \\ 71* \\ 69 \\ 67* \\ 65 \\ 63 \\ 61* \\ 59* \\ 57 \\ 55 \\ 53* \end{pmatrix} \rightarrow$$

$$\rightarrow A_{52}* = A'_{52} \cap {}^*A''_{52} = \begin{pmatrix} 3* & 101* \\ 7* & 97* \\ 31* & 73* \\ 37* & 67* \\ 43* & 61* \end{pmatrix}$$

Finally for the first some even numbers we have:

$$A_2{}^* = \begin{pmatrix} 2* & 2* \end{pmatrix}, \ A_3{}^* = \begin{pmatrix} 3* & 3* \end{pmatrix}, \ A_4{}^* = \begin{pmatrix} 3* & 5* \end{pmatrix}, \ A_5{}^* = \begin{pmatrix} 3* & 7* \\ 5* & 5* \end{pmatrix}, \ A_6{}^* = \begin{pmatrix} 5* & 7* \end{pmatrix},$$

$$A_7{}^* = \begin{pmatrix} 3* & 11* \\ 7* & 7* \end{pmatrix}, A_8{}^* = \begin{pmatrix} 3* & 13* \\ 5* & 11* \end{pmatrix}, A_9{}^* = \begin{pmatrix} 5* & 13* \\ 7* & 11* \end{pmatrix}, \ A_{10}{}^* = \begin{pmatrix} 3* & 17* \\ 7* & 13* \end{pmatrix}, \ A_{11}{}^* = \begin{pmatrix} 3* & 19* \\ 5* & 17* \\ 11* & 11* \end{pmatrix}$$

$$A_{12}{}^* = \begin{pmatrix} 5* & 19* \\ 7* & 17* \\ 11* & 13* \end{pmatrix}, \ A_{13}{}^* = \begin{pmatrix} 3* & 23* \\ 7* & 19* \\ 13* & 13* \end{pmatrix}, \ A_{14}{}^* = \begin{pmatrix} 5* & 23* \\ 11* & 17* \end{pmatrix}, \ A_{15}{}^* = \begin{pmatrix} 7* & 23* \\ 11* & 19* \\ 13* & 17* \end{pmatrix},$$

$$A_{16}{}^* = \begin{pmatrix} 3* & 29* \\ 13* & 19* \end{pmatrix}, \ A_{17}{}^* = \begin{pmatrix} 3* & 31* \\ 5* & 29* \\ 11* & 23* \\ 17* & 17* \end{pmatrix}, \ A_{18}{}^* = \begin{pmatrix} 5* & 31* \\ 7* & 29* \\ 13* & 23* \\ 17* & 19* \end{pmatrix}, \ A_{19}{}^* = \begin{pmatrix} 7* & 31* \\ 19* & 19* \end{pmatrix},$$

$$A_{20}{}^* = \begin{pmatrix} 3* & 37* \\ 11* & 29* \\ 17* & 23* \end{pmatrix}, \ A_{21}{}^* = \begin{pmatrix} 5* & 37* \\ 11* & 31* \\ 13* & 29* \\ 19* & 23* \end{pmatrix}, \ A_{22}{}^* = \begin{pmatrix} 3* & 41* \\ 7* & 37* \\ 13* & 31* \end{pmatrix}, \ A_{23}{}^* = \begin{pmatrix} 3* & 43* \\ 5* & 41* \\ 17* & 29* \\ 23* & 23* \end{pmatrix}$$

$$A_{24}{}^* = \begin{pmatrix} 5* & 43* \\ 7* & 41* \\ 11* & 37* \\ 17* & 31* \\ 19* & 29* \end{pmatrix}, \ A_{25}{}^* = \begin{pmatrix} 3* & 47* \\ 7* & 43* \\ 13* & 37* \\ 19* & 31* \end{pmatrix}, \ A_{26}{}^* = \begin{pmatrix} 5* & 47* \\ 11* & 41* \\ 23* & 29* \end{pmatrix}$$

$$A_{27}{}^* = \begin{pmatrix} 7* & 47* \\ 11* & 43* \\ 13* & 41* \\ 17* & 37* \\ 23* & 31* \end{pmatrix}, \ A_{28}{}^* = \begin{pmatrix} 3* & 53* \\ 13* & 43* \\ 19* & 37* \end{pmatrix}, \ A_{29}{}^* = \begin{pmatrix} 5* & 53* \\ 11* & 47* \\ 17* & 41* \\ 29* & 29* \end{pmatrix}, \ A_{30}{}^* = \begin{pmatrix} 7* & 53* \\ 13* & 47* \\ 17* & 43* \\ 19* & 41* \\ 23* & 37* \\ 29* & 31* \end{pmatrix}$$

$$, A_{31}^* = \begin{pmatrix} 3* & 59* \\ 16*43* \\ 31*31* \end{pmatrix}, \ A_{32}^* = \begin{pmatrix} 3* & 61* \\ 5* & 59* \\ 11* & 53* \\ 17* & 47* \\ 23* & 41* \end{pmatrix}, \ A_{33}^* = \begin{pmatrix} 5* & 61* \\ 7* & 59* \\ 13*53* \\ 19*47* \\ 23*43* \\ 29*37* \end{pmatrix}, \ A_{34}^* = \begin{pmatrix} 7* & 61* \\ 31* & 37* \end{pmatrix}$$

$$A_{35}^* = \begin{pmatrix} 3* & 67* \\ 11* & 59* \\ 17*53* \\ 23*47* \\ 29*41* \end{pmatrix}, \ A_{36} = \begin{pmatrix} 5* & 67* \\ 11* & 61* \\ 13* & 59* \\ 19* & 53* \\ 29* & 43* \\ 31* & 41* \end{pmatrix}, \ A_{37}^* = \begin{pmatrix} 7* & 67* \\ 13* & 61* \\ 31* & 43* \\ 37* & 37* \end{pmatrix}, \ A_{38}^* = \begin{pmatrix} 3* & 73* \\ 5* & 71* \\ 17* & 59* \\ 23* & 53* \\ 29* & 47* \end{pmatrix}$$

$$A_{39}^* = \begin{pmatrix} 5* & 73* \\ 7* & 71* \\ 11* & 67* \\ 17* & 61* \\ 19* & 59* \\ 31* & 47* \\ 37* & 41* \end{pmatrix}, \ A_{40}^* = \begin{pmatrix} 7* & 73* \\ 13* & 67* \\ 19* & 61* \\ 37* & 43* \end{pmatrix} \ A_{41}^* = \begin{pmatrix} 3* & 79* \\ 11* & 71* \\ 23* & 59* \\ 29* & 53* \\ 41* & 41* \end{pmatrix}, A_{42}^* = \begin{pmatrix} 5* & 79* \\ 13* & 71* \\ 11* & 73* \\ 23* & 61* \\ 31* & 53* \\ 37* & 47* \\ 41* & 43* \end{pmatrix},$$

$$A_{43}^* = \begin{pmatrix} 3* & 83* \\ 7* & 79* \\ 13* & 73* \\ 19* & 67* \\ 43* & 43* \end{pmatrix}, \ A_{44}^* = \begin{pmatrix} 5* & 83* \\ 17* & 71* \\ 29* & 59* \\ 41* & 47* \end{pmatrix}, \ A_{45}^* = \begin{pmatrix} 7* & 83* \\ 11* & 79* \\ 17* & 73* \\ 19* & 71* \\ 23* & 67* \\ 29* & 61* \\ 31* & 59* \\ 37* & 53* \\ 43* & 47* \end{pmatrix} \ A_{46}^* = \begin{pmatrix} 3* & 89* \\ 13* & 79* \\ 19* & 73* \\ 31* & 61* \end{pmatrix}$$

$$A_{47}^* = \begin{pmatrix} 5* & 89* \\ 11* & 83* \\ 23* & 71* \\ 41* & 53* \\ 47* & 47* \end{pmatrix},$$

$$A_{48}* = \begin{pmatrix} 7* & 89* \\ 13* & 83* \\ 17* & 79* \\ 23* & 73* \\ 29* & 67* \\ 37* & 59* \\ 43* & 53* \end{pmatrix}, \quad A_{49}* = \begin{pmatrix} 19* & 79* \\ 31* & 67* \\ 37* & 61* \end{pmatrix}, \quad A_{50}* = \begin{pmatrix} 3* & 97* \\ 11* & 89* \\ 17* & 83* \\ 29* & 71* \\ 41* & 59* \\ 47* & 53* \end{pmatrix}, \quad A_{51}* = \begin{pmatrix} 5* & 97* \\ 13* & 89* \\ 19* & 83* \\ 23* & 79* \\ 29* & 73* \\ 31* & 71* \\ 41* & 61* \\ 43* & 59* \end{pmatrix},$$

$$A_{52}* = \begin{pmatrix} 3* & 101* \\ 7* & 97* \\ 31* & 73* \\ 37* & 67* \\ 43* & 61* \end{pmatrix}, \quad A_{53} = \begin{pmatrix} 3* & 103* \\ 5* & 101* \\ 17* & 89* \\ 23* & 83* \\ 47* & 59* \\ 53* & 53* \end{pmatrix}, \quad A_{54} = \begin{pmatrix} 5* & 103* \\ 7* & 101* \\ 11* & 97* \\ 19* & 89* \\ 29* & 79* \\ 37* & 71* \\ 41* & 67* \\ 47* & 61* \end{pmatrix}., \quad A_{55} = \begin{pmatrix} 3* & 107* \\ 7* & 103* \\ 13* & 97* \\ 31* & 79* \\ 37* & 73* \\ 43* & 67* \end{pmatrix}$$

$$A_{56} = \begin{pmatrix} 3* & 109* \\ 5* & 107* \\ 11* & 101* \\ 23* & 89* \\ 29* & 83* \\ 41* & 71* \\ 53* & 59* \end{pmatrix}, \quad A_{57} = \begin{pmatrix} 5* & 109* \\ 7* & 107* \\ 11* & 103* \\ 13* & 101* \\ 17* & 97* \\ 31* & 83* \\ 41* & 73* \\ 43* & 71* \\ 47* & 67* \\ 53* & 61* \end{pmatrix}, \quad A_{58} = \begin{pmatrix} 3* & 113* \\ 7* & 109* \\ 13* & 103* \\ 19* & 97* \\ 37* & 79* \\ 43* & 73* \end{pmatrix}, \quad A_{59} = \begin{pmatrix} 5* & 113* \\ 11* & 107* \\ 17* & 101* \\ 29* & 89* \\ 47* & 71* \end{pmatrix},$$

$$A_{60} = \begin{pmatrix} 7* & 113* \\ 17* & 103* \\ 19* & 101* \\ 23* & 97* \\ 31* & 89* \\ 37* & 83* \\ 41* & 79* \\ 47* & 73* \\ 53* & 67* \end{pmatrix}.$$

HELIX LINES AND RIEMANN'S HYPOTESIS

Tanya Kirilova Mincheva
Sofia-Bulgaria
E-mail:tmintcheva@yahoo.com

Abstract. In this topic we consider the Riemann hypothesis by analyzing some helix lines.

Let us consider a circle with radius one and centre in the beginning of coordinate system in its **real** and **complex equivalents**. Namely:

$$\sigma = cos\theta, \ t = sin\theta$$

or in complex form

$$s = \sigma + it = cos\theta + isin\theta = e^{iQ}$$

We know that each circumference is orthogonal projection of a cylindrical surface. The unit circle is also such a projection, for which the corresponding cylindrical surface intersects the coordinate plane Oxy. This is true because the unit circle intersects the axis Ox in two different symmetrical points. Consequently, the cylindrical surface intersects the plane Oxz in two straight lines, which are parallel to the axis Oz. If a point lies on one of these two lines, then this projection is one of the intersection points on the unit circle with the axis Ox.

If a curve is lying on the cylindrical surface of the unit circle, it will also be projected onto the whole circle or part of it.

After this introduction, let us consider the spatial curve

(1)
$$\sigma = cos\theta,$$
$$t = sin\theta$$
$$z = \vartheta$$

Or as the complex equivalent

(2)
$$s = e^{iQ}$$
$$z = \vartheta$$

The names σ, t, s are chosen out of respect for the memory of Riemann.
In total:

(3)
$$\sigma = \rho cos\theta,$$
$$t = \rho sin\theta$$
$$z = \vartheta$$

Or

(4)
$$s = \rho e^{iQ}$$
$$z = \vartheta$$

The equations from (1) to (4) we call equations of **helix lines**. (See [3], [5])
They are well known in the textbooks on differential geometry. For example [3], [7].

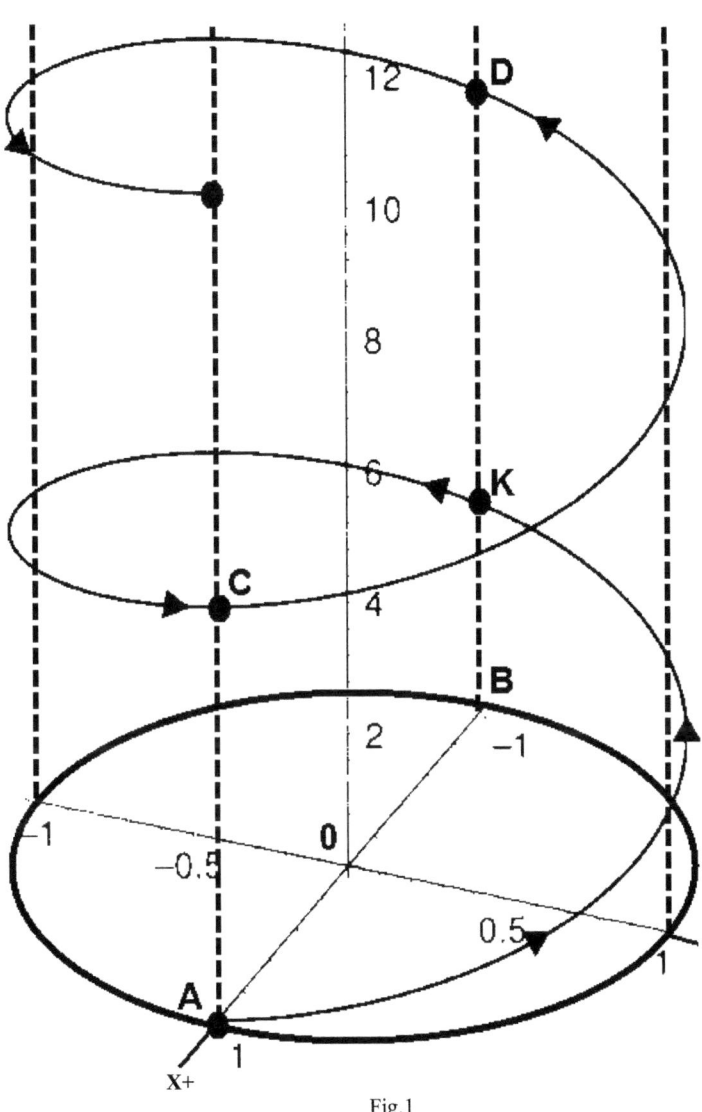

Fig.1

This figure is a copy from the Internet. [See "Helix"]. The helix f= (cos(t), sint(t), t) from t=0 to 4π, with arrowheads showing the direction of the increasing f. The depiction of its projection as an ellipse and the fixed points on the helix which are projected onto the axis are introduced by the author.

Geometric analysis.
From the analytical expressions in (1)-(4) and their corresponding graphs, we understand, that all points, where helix lines intersect with the plane **OXZ** are projected in two points from the axis **OX**. These are the points where the circumference, which is a projection of the cylindrical space, intersects the axis-**Ox**. *In other words, they are nontrivial zeros for the circumference-function in the plane OXY.* The coefficient *b* of the third equality defines the step of the helix line. We know that all helix lines on a cylindrical surface are projected onto the same circle. For this reason, if we have to define one helix line from a unit circle, in the expression for **Z** we choose coefficient 1. *Finally, all different points in which the helix line intersects the plane OXZ are projected on the X-axis in nontrivial zeros for the respective circumference.* All these considerations are simple and well known as a cylindrical surface. See [3], [5], [7] and etc.

Theorem 1. Let

$$F(s) = n^{-s}, \quad n \in N.$$

Be a function of complex variable $s = \sigma + it$. Then

$$R_e[F(s)] = n^{-\sigma} \cos(\ln nn^{-t}), \quad I_m[F(s)] = n^{-\sigma} \sin(\ln n \, n^{-t}), \quad n \in N .$$

Proof.

(5) $$F(s) = n^{-s} = e^{\ln n^{-\sigma - it}} = n^{-\sigma} e^{i \ln n^{-t}} = n^{-\sigma} \cos(\ln n^{-t}) + in^{-\sigma} \sin(\ln n^{-t})$$

The strong proof has the following key moments. (See [4], [5], [6])

$$F(s) = F(\sigma + it) = U(\sigma, t) + iV(\sigma, t).$$

We shall use the following property of the function n^{-s} :
I. $n^{-s} = n^{-\sigma - it} = n^{-\sigma} . n^{-it}$.

II. $n^{-s} = n^{-\sigma - it} = n^{-\sigma}$ on the real axis, when in complex plane $y = 0$.

III. n^{-s} satisfies the Cauchy-Riemann conditions.

$U(\sigma, t) = \text{Re}(n^{-\sigma - it}) = n^{-\sigma} \text{Re}(n^{-it})$. If we denote $\text{Re}(n^{-it}) = u(t)$, then

$$U(\sigma, t) = n^{-\sigma} u(t).$$

Tanya Mincheva

$V(\sigma,t)=\mathrm{Im}(n^{-\sigma-it})=n^{-\sigma}\,\mathrm{Im}(n^{-it})$. If we denote $\mathrm{Im}(n^{-it})=v(t)$, then

$$V(\sigma,t)=n^{-\sigma}v(t).$$

Using the equations of Cauchy-Riemann we have

$$\left| \frac{\partial U}{\partial \sigma}=\frac{\partial V}{\partial t} \right.$$

$$\left| \frac{\partial V}{\partial \sigma}=-\frac{\partial U}{\partial t} \right.$$

Or if we denote $l_n n^{-1}=p$ we have the system

$$\left| pu(t)=\frac{dv}{dt} \right.$$

$$\left| pv(t)=-\frac{du}{dt} \right.$$

And in the process of resolving we get

$$\left| p^2u(t)+u''(t)=0 \right.$$

$$\left| p^2v(t)+u''(t)=0 \right.$$

Hence

$$\left| \begin{array}{l} u(t)=A\cos(pt)+B\sin(pt) \\ v(t)=C\cos(pt)+D\sin(pt) \end{array} \right.$$

A=?, B=?, C=?, D=?.

$t=0 \to u(0)=A\cos(0)+B\sin(0) \to$
$A=u(0)$, but
$u(0)=\mathrm{Re}(n^0)=1 \to$
$\to A=1$
Analogy:
$v(0)=\mathrm{Im}(n^{-i.0})=0 \to$
$\to C=0$
So far we have obtained

$$u(t) = \cos(pt) + B\sin(pt)$$
$$v(t) = D\sin(pt)$$ and

From the system …using these last values of $u(t)$ and $v(t)$ we have

$$\cos(pt) + B\sin(pt) = D\cos(pt) \to D = 1$$
$$0 = B\cos(pt) \to B = 0$$
Or $u(t) = \cos(pt)$ and $v(t) = \sin(pt)$
Finally

$$F(s) = F(\sigma + it) = n^{-\sigma}\cos(\ln n^{-t}) + in^{-\sigma}\sin(\ln n^{-t}).$$
That finishes the proof.

Theorem 2. Let

$$\vec{r}_n = (r_{n1}, r_{n2}, r_{n3}), \text{ where } r_{n1} = n^{-\sigma}\cos(\ln n^{-t}), r_{n2} = n^{-\sigma}\sin(\ln n^{-t}), r_{n3} = b_n \ln n^{-t}.$$

be a vector-functions of σ and t for every $n \in N$, σ and t, σ is a constant.
We will prove, that the necessary and sufficient condition for these three-dimensional functions to be a helix for every $n \in N$, is $\sigma = 1/2$.
Remark: This function has and a complex analogue, if we use Euler's formula [See(5)].

Proof.

For successive derivatives of t we obtain:

$$\vec{r}_n' = \left(-n^{-\sigma}\sin(\ln n^{-t})\ln n^{-1}, \ n^{-\sigma}\cos(\ln n^{-t})\ln n^{-1}, \ b_n \ln n^{-1}\right)$$

$$\vec{r}_n'' = \left(-n^{-\sigma}\cos(\ln n^{-t})(\ln n^{-1})^2, \ -n^{-\sigma}\sin(\ln n^{-t})(\ln n^{-1})^2, \ 0\right)$$

$$\vec{r}_n''' = \left(n^{-\sigma}\sin(\ln n^{-t})(\ln n^{-1})^3, \ -n^{-\sigma}\cos(\ln n^{-t})(\ln n^{-1})^3, \ 0\right)$$

$$\left|\vec{r}_n' \times \vec{r}_n''\right| = \sqrt{n^{-2\sigma}b_n^2(\ln n^{-1})^6 + n^{-4\sigma}(\ln n^{-1})^6} = n^{-\sigma}(\ln n^{-1})^3\sqrt{b_n^2 + n^{-2\sigma}}$$

$$\vec{r}_n'\vec{r}_n''\vec{r}_n''' = n^{-2\sigma}b_n^2(\ln n^{-1})^6$$

$$\left|\vec{r}_n'\right|^3 = \left(\sqrt{n^{-2\sigma}(\ln n^{-1})^2 + b_n^2(\ln n^{-1})^2}\right)^3 = (\ln n^{-1})^3(n^{-2\sigma} + b_n^2)^{3/2}$$

$$\left(\vec{r}_n{}' \times \vec{r}_n{}''\right)^2 = n^{-2\sigma} b_n^2 \left(\ln n^{-1}\right)^6 + n^{-4\sigma}\left(\ln n^{-1}\right)^6 = n^{-2\sigma}\left(\ln n^{-1}\right)^6\left(b_n^2 + n^{-2\sigma}\right)$$

Hence from

$$\kappa_n = \frac{\left|\vec{r}_n{}' \times \vec{r}_n{}''\right|}{\left|\vec{r}_n{}'\right|^3} \quad \text{and} \quad \tau_n = \frac{\vec{r}_n{}'\,\vec{r}_n{}''\,\vec{r}_n{}'''}{\left(\vec{r}_n{}' \times \vec{r}_n{}''\right)^2} \quad \text{we have finally}$$

$$\kappa_n = \frac{n^{-\sigma}\left(\ln n^{-1}\right)^3\left(b_n^2 + n^{-2\sigma}\right)^{1/2}}{\left(\ln n^{-1}\right)^3\left(b_n^2 + n^{-2\sigma}\right)^{3/2}} = \frac{n^{-2\sigma}}{\left(b_n^2 + n^{-2\sigma}\right)} =$$

$$\frac{n^{-2\sigma}}{n^{-2\sigma}\left(b_n^2 n^{2\sigma} + 1\right)} = \frac{1}{b_n^2 n^{2\sigma} + 1}$$

$$\tau_n = \frac{\vec{r}_n{}'\,\vec{r}_n{}''\,\vec{r}_n{}'''}{\left(\vec{r}_n{}' \times \vec{r}_n{}''\right)^2} = \frac{n^{-2\sigma}\left(\ln n^{-1}\right)^{-6} b_n^2}{n^{-2\sigma}\left(\ln n^{-1}\right)^{-6}\left(b_n^2 + n^{-2\sigma}\right)} = \frac{b_n^2}{b_n^2 + n^{-2\sigma}}$$

Or

$$\tau_n = \frac{b_n^2 n^{2\sigma}}{b_n^2 n^{2\sigma} + 1}$$

$$\frac{\tau_n}{\kappa_n} = b_n^2 n^{2\sigma}$$

This denotes, that

$$\tau_n = \frac{b_n^2 n^{2\sigma}}{b_n^2 n^{2\sigma} + 1} \text{ and } \kappa_n = \frac{1}{b_n^2 n^{2\sigma} + 1} \text{ are functional dependants for every } n \in N.$$

Let us consider the following functions:

$$\tau(x,\sigma) = \frac{b_n^2 x^{2\sigma}}{b_n^2 x^{2\sigma} + 1}, \quad \kappa(x,\sigma) = \frac{1}{b_n^2 x^{2\sigma} + 1}$$

Obviously for $x = n$ we have $\tau = \tau_n, \kappa = \kappa_n$. The functional dependence between the last two functions is:

$$\tau = b_n^2 x^{2\sigma} \kappa$$

$$f(x) = \frac{\tau(x,\sigma)}{\kappa(x,\sigma)} = b_n^2 x^{2\sigma} \text{ for every } x.$$

For every x we have a parameter σ.

$$f'(x) = \left(\frac{\tau(x,\sigma)}{\kappa(x,\sigma)} \right)' . \leftrightarrow$$

$$\leftrightarrow f'(x) = b_n^2 \, 2\sigma x^{2\sigma-1}$$

From the condition "for every x"

$$2\sigma - 1 = 0.$$

Hence

$$\sigma = \frac{1}{2}.$$

That is for every x, hence and for every $n \in N$
Hence

$$\tau_n = \frac{b_n^2 n}{b_n^2 n + 1}, \quad \kappa_n = \frac{1}{b_n^2 n + 1}$$

Therefore, for each fixed **n** they are constants.
The sufficient condition is elementary.

This finishes the proof.

Riemann's Hypothesis. The real part of every non-trivial zero of the Riemann's zeta function

$$\varsigma(s) = \sum_{n=1}^{\infty} \frac{1}{n^s} = \frac{1}{1^s} + \frac{1}{2^s} + \frac{1}{3^s} + \ldots + \frac{1}{n^s} + \ldots$$

is ½
Proof.

Using **T1** we have

$$\varsigma(s) = \sum_{n=1}^{\infty} \left\{ \left(n^{-\sigma} \left(\cos \ln n^{-t} + i \sin \ln n^{-t} \right) \right) \right\}$$

There $s = \sigma + i\tau$. For each **n** we complement $n^{-\sigma} \left(\cos \ln n^{-t} + i \sin \ln n^{-t} \right)$ to three-dimensional space curve lines of the real equivalents $\vec{r}_n = (r_{n1}, r_{n2}, r_{n3})$, where

$$r_{n1} = n^{-\sigma} \cos\left(\ln n^{-t} \right), r_{n2} = n^{-\sigma} \sin\left(\ln n^{-t} \right), r_{n3} = b_n \ln n^{-t}.$$

Using **T2** these three-dimensional lines are helix lines only for $\sigma = 1/2$.
We know from our **Geometric analysis** that all different points in which the helix line intersects the plane **OXZ** are projected on the **X**-axis in nontrivial zeros for the respective circumference.
This completes the proof of the Riemann hypothesis.

Conclusion. The common thing in both topics is that we examine multitudes, the elements of which are also multitudes.

References:

1. Chakalov L.-Theory of Differential Equations-(Bulgarian), Sofia 1960.
2. ShmelevP.- Theory of series in exercises and tasks-Moscow-1983.(Russian.)
3. Vuigodski M.- "Differential geometry"-Moscow-1949.(Russian).
4. Shahno K.-"Elements of theory of complex functions and operational calculus"-Minsk-1975.(In Russian)
5. Internet: "Helix"-The free encyclopedia.(English)
6. Internet: "Complex Analysis" The free encyclopedia.(English).
7. Internet: "Differential geometry" The free encyclopedia.(English)
8. Internet: "Riemann Hypothesis"-Wikipedia, the free encyclopedia.(English) Etc.

ABOUT ONE CORRECTION OF THE METHOD OF
"THE IDEAL POINT" IN EUCLIDEAN METRIC

Tanya Kirilova Mincheva

Sofia-Bulgaria

E-mail:tmintcheva@yahoo.com

Abstract. In the presented work is proved the necessity by a correction of the method of "the ideal point "in theory of the vector optimization and is made such in Euclidean metric. We consider and an economical interpretation, because she is very important for a large numbers specialists in different areas.

In theory of the vector optimization one a most important problem is the problem for find out only one alternative, the best in some sense, by all the rest alternatives. In some cases it is sufficiently only to find Parreto-optimal alternatives.

Let we have p real functions:

$$(1) \qquad f_1(x), \dots, f_p(x); \qquad x \in X \subseteq R^n,$$

what are criteria for quality of alternatives x from multitude $X \subseteq R^n$. We call that the alternative x_1 dominates by Parreto the alternative x_2 $(x_1 \succ x_2)$ if

$$(2) \qquad f_i(x_1) \ge f_i(x_2); \quad i = 1 \div p; \quad x_1, x_2 \in X \subseteq R^n,$$

where at least from those inequalities is a strong inequality.

The exact definition is the following $[1]$:

The alternative x^* is a Parreto optimal alternative for the criteria (1) if we have simultaneously:

$$(3) \qquad f_i(x^*) \ge f_i(x) \quad i = 1 \div p; \quad x, x^* \in X \subseteq R^n,$$

and for every another alternative, for which are possible (3), they are possible only as an equalities.

On every alternative $x \in X \subseteq R^n$ corresponds her vector valuation:

$$(3) \qquad f_1(x), \dots, f_p(x)$$

When x describes the multitude $X \subseteq R^n$, then in the multitude of the vector valuations R^p we obtain the multitude F, which we call a multitude of the vector values. By means of Parreto optimal alternatives πx we obtain Parreto–optimal vector values πF .

Examples:

I. $f_1(x) = -x^2 + 6x - 5$

$f_2(x) = -0.5x^2 + 3x + 3.5$

$f_3(x) = -x^2 + 6x$

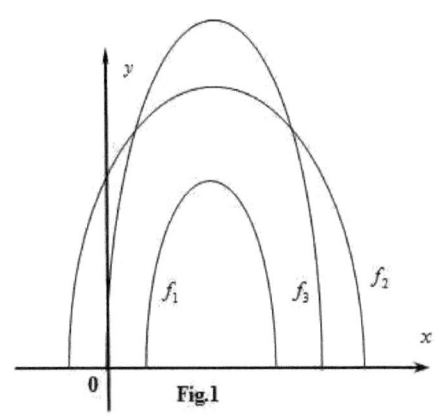

Fig.1

II. $f_1(x) = \begin{cases} -x^2 + 2x + 3 & \text{if } 0 \le x < \\ 4 & \text{if } 1 \le x < \\ -x^2 + 8x - 12 & \text{if } 4 \le x < \end{cases}$

$f_2(x) = -x^2 + 4 \text{ if } 0 \le x \le 8$

$f_3(x) = -(1/4)x^2 + 2x \text{ if } 0 \le x \le 8$

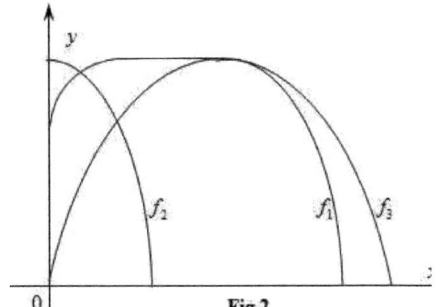

Fig.2

III. $f_1(x) = x + 1 \quad \text{if } 0 \le x \le$

$f_2(x) = -x + 3 \quad \text{if } 0 \le x \le$

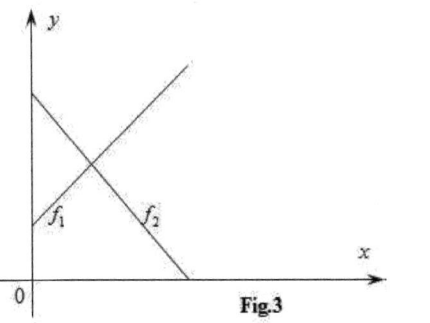

Fig.3

In the first example $x_0 = 3$ is only one Parreto–optimal alternative. There she is a point of the absolute maximum. That solves the problem for to find only one alternative better from another:

55

This is $x_0 = 3$, for which the three functions have a maximum. If these functions are mould the search of three merchandises simultaneously, then $x_0 = 3$ define the situation for with this search is too big for the three articles. For the second and the third examples this is not true.

In the second example the criteria are contradictory in the interval [1, 4). In the third example the criteria are contradictory in every interval of the axis O_x.

Despite of all that the problem for fined out only one alternative remains free of answer and his practical value is indisputable.

The last investigations consist of tree kind methods: a priori, a posteriori and adaptable. One from the a priori researches is a "method of the ideal point in the space of vector values". The problem is the following:

In the space of vector values R^p we consider the point:

$$(5) \qquad y_0 = (y_{10}, y_{20}, ..., y_{p0})$$

where

$$(6) \qquad y_{i0} = \max f_i(x) = f_i(x_{i0}); \ x \in X \subseteq R^n$$

and we suppose that the multitude of vector values is limited and closed. If

$$(7) \qquad x_{10} = x_{20} = ... = x_{p0} = x_0$$

then obviously x_0 is the best alternative. Namely this is "the ideal point". Such is the case in our first example: $x_0 = 3$. But this situation is possible in exceptional rare cases. Then is logically to consider the alternative, which is nearest to "the ideal point". In the works until now researches propound for that sort of point to be the point for which the vector value is nearest to the vector value of "the ideal point". This denote that for the function

$$(8) \qquad y = F(x) = (f_1(x), f_2(x), ..., f_p(x)), \quad \text{i.e.} \quad F(x): R^n \to R^p$$

we must to solve the following extremely problem

$$(9) \qquad \rho(y_0, y) = \inf_{x \in X} \left(\sum_{i=1}^{p} (f_i(x) - f_i(x_{i0}))^2 \right)^{1/2}$$

There are known the defects of this method when we using (9). For a better and convincing elucidation to the correction of the formula (8) – (9) and in compliance with the our problem we cite directly from [1] p.46- 47 these criticisms…"When we speak for find out a point "the most near by the ideal point" we suppose introduced a metric in the space of vector values…but a metrics in

R^n we can introduce by different modes. Obviously for different metrics, the best will be the different alternatives. But we can prove that for every π-optimal alternative, we can find out a metric where the vector value of this alternative is nearness for "ideal point". Hence the problem for "ideal point" is a problem for choice to metric, what is not simple problem.

Even though if we have one good metric, which corresponds to the character of a practical problem that yet not solves the problem. Really the function distance to "the ideal point" give a possibility to set in order the all alternatives and to choose the best alternative. Arises the question how reasonable is that selection from the point of view of the theory of the rational choice.

One of the axioms in this theory is the axiom for independence by nonparticipate alternatives. As per that axiom the result by compare of the alternatives **a** and **b** depend only on them and independs by the presence or the absence of someone third alternative **c**.

It is easily to see that the method of "the ideal point" non-satisfies the axiom for the independence. Namely, let vector value f(x') is more near to "the ideal point" than f(x''). Hence x' is more good by x". But if we add to the multitude X a point x*, then we can create such displacement from "the ideal point", that to obtain x" as a more good alternative by x' (Fig.4)

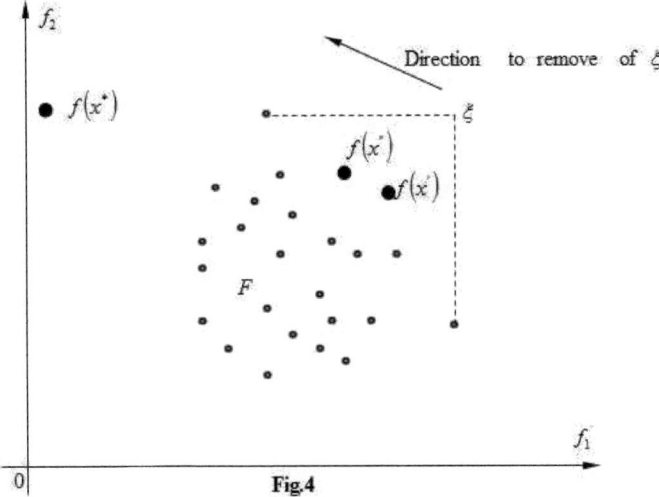

Fig.4

But there exist very many situations, which not satisfy the axiom for independence and though they are intuitively acceptable.

For example let we have as a multitude of alternatives a convex plane disc, and let our problem is to find a supporting point, for to insure a horizontally position of the figure. Obviously this is the center of gravity and the rests are set in order accordingly the distance by the center of gravity. But this alternative unsatisfied the axiom for independence by nonparticipants alternatives.

Till we unfound theoretical situations which enable with very good precision to determinate the possible principles of optimization for answer to concrete practical problems by then the possibilities for application of these procedures are small." (End of the citations)

I.e. we have not a complete theory for solve some of multi-criteria problems. For that reason we want to make one correction of this method, i.e. of the formula (9). For completeness of our statement we remind some definitions and theorems from the Functional and the Real analysis witch have natural contacts whit our problem.

Definition 1 [4]: *Let E be a linear space and in E be two different means are defined the norms:* $\|x\|^1$ *and* $\|x\|^2$. *We call two norms equivalents, if exist numbers $\alpha > 0$ and $\beta > 0$ such that for every $x \in E$ we have* $\alpha\|x\|^1 \leq \|x\|^2 \leq \beta\|x\|^1$.

I.e. the concept equivalent has the characteristics reflexibility, symmetrically and transitivity.

Theorem 1 [4] *In every linear finite – measure space everybody norms are equivalents.*

Definition 2 [4] *A space with scalar multiplication we call Halberd's space if it is fullness about norm, which correspond of that scalar multiplication.*

The simplest example of Halberd's space this is the Euclidean space. There scalar multiplication we define by

$$\langle x, y \rangle = \sum_{i=1}^{n} x_i y_i$$

The respective norm and metric are

$$\|x\| = \left(\sum_{i=1}^{n} x_i^2\right)^{1/2}; \qquad \rho(x,y) = \|x - y\| = \left(\sum_{i=1}^{n} (x_i - y_i)^2\right)^{1/2}$$

Definition 3. [4] *Let x_0 is any point and A is nonempty multitude by points. The distance from x_0 to the multitude A we define by*

$$\rho(x_0, A) = \inf_{y \in A} \rho(x_0, y)$$

Definition 4. [4] *Let A and B are two nonempty point-multitudes. Then the distance from A to B we define by*

$$\rho(A, B) = \inf_{x \in A, y \in B} \rho(x, y)$$

From this definition immediately follows, that $\rho(A, B)$ always exists and always $\rho(A, B) \geq 0$.

Theorem 2. $[4]$ *Let A and B are two nonempty multitudes, which are closed and at least from them is limited. Then exists points $x^* \in A, y^* \in B$ such that*

$$\rho(x^*, y^*) = \rho(A, B)$$

Consequences:

I. *If A and B are closed multitudes such that at least from them are limited and $\rho(A, B) = 0$, then*
$$A \cap B \neq \varnothing.$$

II. *Let x_0 be any point and F is nonempty closed multitude. Then in F exists point x^*, such that*

$$\rho(x_0, x^*) = \rho(x_0, F)$$

III. *If for the point x_0 and for the closed multitude F we have $\rho(x_0, F) = 0$, then $x_0 \in F$.*

For our problem the multitude A consists a finite number points. Namely let $f_1(x), ..., f_p(x)$ are p real functions in the compact multitude $X \subset R^n$. Let by B we denote the multitude
$$B = \{(x, f_1(x)), (, f_2(x)), ..., (, f_p(x))\}$$
If denote

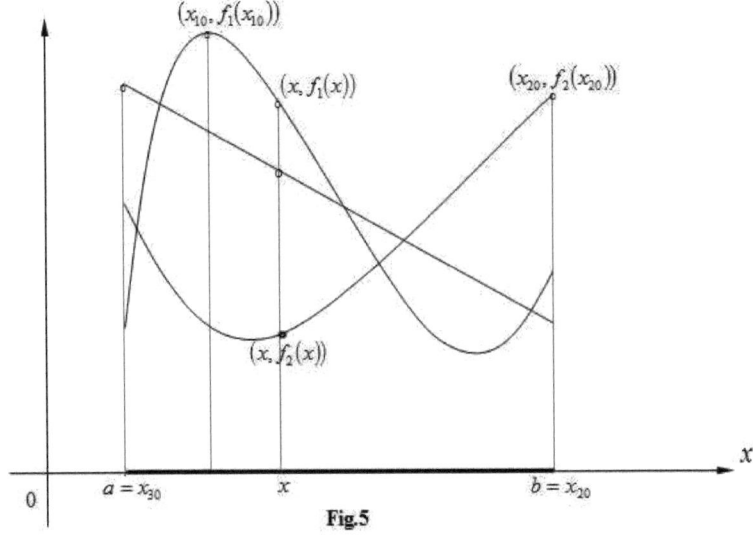

$$x_{i0} = \arg\sup_{x \in X} f_i(x) \quad x_{i0} \in X \subset R^n \quad i = 1, ..., p, \quad f_{i0} = \sup_{x \in X} f_i(x),$$

then we can define the discrete multitude(Fig.5):

$$A = \{(x_{10}, f(x_{10})), (x_{20}, f(x_{20})), ..., (x_{p0}, f(x_{p0}))\}$$

In the sense of Euclidean metric the distance $\rho(A, B)$ is

$$\rho(A, B) = \inf_{x \in B} \left\{ \sum_{i=1}^{p} \left[(f_i(x) - f_{i0})^2 + (x - x_{20})^2 \right] \right\}^{1/2}$$

According to T2 there exists point $x^* \in X \subset R^n$ such that

$$\rho(A, B) = \inf_{x \in X} \left\{ \sum_{i=1}^{p} \left[(f_i(x) - f_{i0})^2 + (x -_{20})^2 \right] \right\}^{1/2} = \left\{ \sum_{i=1}^{p} \left[(f_i(x^*) - f_{i0})^2 + (x^* - x_{20})^2 \right] \right\}^{1/2}$$

Theorem 3. *For the given p scalar functions* $f_1(x), \ldots, f_p(x)$, $x \in X \subset R^n$ *the point* $x^* \in X$ *is "ideal point" then and only then, when x* satisfy the condition:*

$$x^* = \arg\inf_{x \in X} \left\{ \sum_{i=1}^{p} \left[(f_i(x) - f_{i0})^2 + (x - x_{20})^2 \right] \right\}^{1/2} ; x \in X \subset R^n \text{ , such that}$$

$$\left\{ \sum_{i=1}^{p} \left[(f_i(x^*) - f_{i0})^2 + (x^* - x_{20})^2 \right] \right\}^{1/2} = 0$$

Proof.

The proof is elementary. Namely, let x^* is "the ideal point". I.e. $x_{10} = x_{20} = \ldots = x_{p0} = x^*$.

Then obviously the conditions in the theorem are satisfied. The contrary conclusion is to verify, because every some of nonnegative numbers is eagle of zero only if every by this numbers is eagle of zero too. This finishes the proof.

After this theorem we have the possibility to formulate the next fundamental for our problem
Definition5. *Let*

$$f_1(x), f_2(x), \ldots, f_p(x) \quad x \in X \subset R^n$$

be p scalar functions, where X is a compact in R^n *alternative* $x^* = (x_1^*, \ldots, x_n^*)$ *is a solution of the problem for to find out of unique alternative for the given p scalar continuous functions then and only then, when x* is a solution of the following extremely problem:*

$$(10) \qquad x^* = \arg\inf_{x \in X} \left\{ \sum_{i=1}^{p} \left[(f_i(x) - f_{i0})^2 + (x - x_{i0})^2 \right] \right\}^{1/2} ; x \in X \subset R^n.$$

Now we must to explain our correction by means of reply to the question how the addition to $\sum_{i=1}^{p} (x - x_{i0})^2$ make better the solution.

I.First of all (10) consists as a trivial particular kind the formula for distance from point to multitude when we consider only one criterion. Namely

$$\rho = \{\ (f(x) - f(x_0))^2 + (x - x_0)^2\ \}^{1/2}.$$

Until only $\left| f(x) - f(x_0) \right|$ is wrong. (Fig.6)

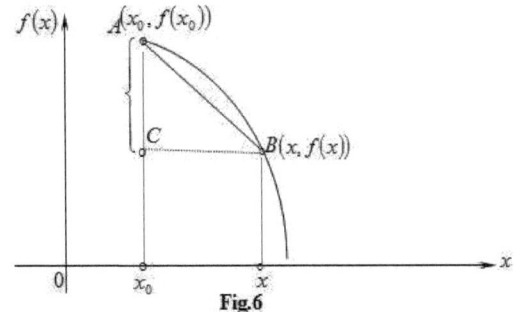

Fig.6

This fact is well known by the elementary mathematics. I.e. $AC \le AB$
This denotes that we shall obtain using (9) on principle erroneous results. Such is the case whit the convex plane disc what we have already considered. There for every admissible alternative x
$f_i(x) - f_i(x) = 0$ i=1,...,p and for that reason we have

$$(11)\quad x^* = \arg\inf_{x \in X} \left\{ \sum_{i=1}^{p} \left[(f_i(x) - f_{i0})^2 + (x - x_{i0})^2 \right] \right\}^{1/2} = \arg\inf_{x \in X} \left\{ \sum_{i=1}^{p} (x - x_{i0})^2 \right\}^{1/2} \quad x \in X \subset R^2$$

In $[1]$ is not explains how by means of (9) we conclude that this point is the center of gravity. By mechanics we know that namely (11) is the formula for the inertia momentum about the center of gravity

II. That last interpretation is exclusively important, because proves that the axiom for independence not every is reasonable. However we know that adding of points, which general moment of inertia are equal to zero, keeping the equilibrium. Hence has sense the question about analysis of the methods, through which we can adding or ignoring alternatives.

.
III. After this correction we can make the following economic interpretation: not only the functions - private criteria $f_1, f_2, ..., f_p$, but and the alternatives x have meaning of criteria. Hence is expedient $\left| f_i(x) - f_{i0} \right|$ to be called "equilibrium prizes" corresponding on the diversions $\left| x^* - x_{i0} \right|$

IV. About to choice of a metrics this question arises not only in theory of vector optimization. For very many problems of the economics, the physics and others the

61

Euclidean space is proved fundament. That not overrides the necessity of analogy researches for another spaces and respective metrics, but we shall investigate in next works.

Remark. We suppose everyone the functions in the same dimensions. In contrary we consider the problem in the following kind:

$$x^* = \arg\inf_{x \in X} \left\{ \sum_{i=1}^{p} \left[\left(\frac{f_i(x) - f_{i0}}{f_{i0}} \right)^2 + \left(\frac{x - x_{i0}}{x_{i0}} \right)^2 \right] \right\}^{1/2} \; ; x \in X \subset R^n.$$

Hence if some of components to (x_{i0}, f_{i0}) are eagle to zero, then is necessary suitable change of the co-ordinates.

Conclusion. With our correction we aim at to create one general description of the some problems in theory of vector optimization. Such is the problem for find out only one alternative the best by all the rest alternatives. We would like to think that this correction be an impulse for future researches in the Calculus of variations and Optimal control theory. Coming is an analogical correction for another metrics and applications for solve many problems in economics, physics, etc. That is prognosis for next publications.

REFERENCES

1.Dubov U.A., Travkin S.I., Iakimec V.N.-"Multi-criteria models to compose and to choose variant
Systems"-Nauka-M.1986 (in Russian)
2. Ilin V.A., Sadovnitchi V.A., Sendoff Bl."Mathematical analysis"-Nauka-M.1979 (In Bulgarian).
3.Hutson V.C.L. and John Pym "Applications of Functional Analysis and Operator Theory" –M 1980 (in Russian).
4.Natanson I.P."Introduction in theory of real functions"-Sofia-1971 (in Bulgarian)

Acknowledgements: I am grateful to the student Hristo Ivanov for the translation and to my son Kiril Minchev for participation in computer layout.

Tanya Kirilova Mincheva
Sofia
Bulgaria
E-mail:tmintcheva@yahoo.com

Printed by Books on Demand GmbH, Norderstedt / Germany